High-Voltage Test Techniques

Dieter Kind
Kurt Feser

2nd Revised and Enlarged Edition

With 211 Figures and 12 Laboratory Experiments

Translated from the German by Y. Narayana Rao
Professor of Electrical Engineering (Retd.)
Indian Institute of Technology, Chennai, India

Newnes

OXFORD BOSTON AUCKLAND JOHANNESBURG MELBOURNE NEW DELHI

Newnes
An imprint of Butterworth-Heinemann Ltd
Linacre House, Jordan Hill, Oxford OX2 8DP
225 Wildwood Avenue, Woburn, MA 01801-2041
A division of Reed Educational and Professional Publishing Ltd

 A member of the Reed Elsevier plc group

First published

British Library Cataloguing in Publication Data
A catalogue record for this book is available from the British Library

ISBN 0 7506 5183 0
Transferred to digital print on demand, 2005
Printed and bound by Antony Rowe Ltd, Eastbourne

From the Preface to the 1st English Edition (in 1978)

High-voltage technology is a field of electrical engineering, the scientific principles of which are essentially found in Physics and which by its application, is intimately linked with industrial practice. It is concerned with the physical phenomena and technical problems associated with high voltages.

The properties of gases and plasmas, as well as liquid and solid insulating materials, are of fundamental significance to high-voltage technology. However, despite all progress, the physical phenomena observed in these media can only be incompletely explained by theoretical treatment, and so experiment constitutes the foreground of scientific research in this field. Teaching and research in high-voltage technology thus rely mainly upon experimental techniques when dealing with problems.

Recognition of this fact is the conceptual basis for the present book. It is primarily intended for students of electrical engineering and aims to provide the reader with the most important tools for the experimental approach to problems in high-voltage technology. An attempt has been made here to indicate important practical problems of testing stations and laboratories, and to suggest solutions. The book should therefore also prove to be a help to the work of the practicing engineer.

The theoretical considerations are correlated with the experiments of a high-voltage practical course, which are described in great detail. The treatment assumes as much familiarity with the subject as may be expected from 3rd year students of electrical engineering.

The development of high-voltage technology reaches as far back as the early years of the 20th century. Meanwhile, numerous new branches of electrical engineering exist of which every electrical engineer must possess some knowledge. This development necessarily led to reconsideration of the common scientific principles of electrical engineering, and this naturally influenced the traditionally based terminology of high-voltage technology. Physical quantities have been given throughout in the international system of units "SI".

As far as 30 years back, my esteemed predecessor Prof. Dr.-Ing. E.h. *Erwin Marx* treated the subject matter of this book in his "Hochspannungspraktikum", a book widely circulated in Germany and abroad. In those days the highest transmission voltage was 220 kV, today overstepping the

1 MV mark is within reach. The fact that in the meantime the development of high-voltage technology has continued in leaps and bounds justifies a thorough revision of the same material.

This is an updated English version of the German book "Einführung in die Hochspannungs-Versuchstechnik", which was published in its first edition in 1972 as a result of long years of experience, both in teaching and research, at the Technische Universität Braunschweig. Numerous colleagues in the High-Voltage Institute of this university made substantial contributions to the contents of the book as well as to the planning and verification of the described experiments. Particular thanks go to Dr. *Walter Steudle* for his revision of the manuscript and to Mr. *Hans-Joachim Müller* for his exemplary preparation of the drawings.

As far as the present English version is concerned, I wish to express my sincere gratitude to my colleague Dr.-Ing. *Narayana Rao*, Indian Institute of Technology, Madras, who carried out the translation as an experienced scientist and engineer. His work was supplemented by Mrs. *C.C.J. Schneider* M.A.(Cantab), who carefully revised the whole manuscript. Thanks are also due to Dr. *Tim Teich*, UMIST, Manchester, for his competent help, and to the publishers Vieweg-Verlag for their understanding readiness to comply with special requests.

The 1972 edition of the book has meanwhile been well received. It is my sincere wish that this English edition may now also become a modest contribution to the progress in high-voltage technology beyond the German speaking countries.

Dieter Kind

Foreword to the 2nd English Edition

The present 2nd English edition of the book with the shortened title "High-Voltage Test Techniques" is a completely revised and enlarged edition of the book "Introduction to High-Voltage Experimental Techniques", which has acquired in many countries a firm place especially in the training of students of Electrical Engineering. The original German edition has been translated into English, Chinese, Turkish, Indonesian and Persian languages. The present revision, undertaken jointly by two authors, takes into account not only the latest international developments in High-Voltage and Measurement Technology, but addresses, more than its predecessor, engineers of the testing field. The authors hope that with this, they would be addressing a still larger circle of interested people.

High-Voltage Technology surely belongs to the traditional areas of Electrical Engineering, but has in no way stood still in its development. New insulating materials, computing methods and voltage levels pose repeatedly new problems or open up methods of solution; electromagnetic compatibility (EMC) of components and systems also demands increased attention. The Authors hope that their experience brought out in this book would be of use to students of Electrical Engineering confronted with high-voltage problems in their studies, in research and development and also in the testing field.

We thank the many colleagues for the support extended by their advice and help in our effort to bring the standard work on experimental high-voltage technology up-to-date and thus have a reference source for study, research and testing field practice.

We thank the publishers Vieweg Verlag and Shankar's Book Agency for their good cooperation.

Braunschweig/Stuttgart
March 1999

Dieter Kind
Kurt Feser

LIST OF CONTENTS

List of symbols used

a	length		L	self inductance
b	width, atmospheric pressure, mobility		L'	inductance per unit length
			M	mutual inductance
c	velocity of light, length		N	number of turns
d	diameter, relative air density		P	power, probability
f	frequency		P'	power density
i	current (instantaneous value), running index		Q	electrical charge, heat quantity
			R	resistance, radius
k	proportionality factor		S	current density, rate of voltage rise (steepness)
l	length			
m	mass, natural number		T	periodic time; time constant, response time
n	natural number, pulse rate, charge carrier density			
			U	voltage (fixed value)
p	pressure, Laplace-operator		U_{rms}	voltage (r.m.s.value)
q	charge		$\bar{U}$	voltage (arithmetic mean value)
r	radius, spacing			
s	gap distance, standard deviation		$\hat{U}$	voltage (peak value)
			W	energy
$s(t)$	step function		W'	energy density
t	time		X	reactance
u	voltage (instantaneous value)		Y	admittance
$\ddot{u}$	open-circuit transformation ratio		Z	surge impedance, apparent resistance
v	velocity, coefficient of variation		α	ionisation coefficient, abbreviation
$w(t)$	unit step response		β	angle
x	local coordinate		δ	loss angle
z	local coordinate		$\tan \delta$	dissipation factor
			ε	dielectric constant, earthing coefficient
A	area, constant			
B	magnetic induction, constant		η	utilisation factor
C	capacitance		ϑ	temperature
C'	capacitance per unit length		χ	conductivity
D	dielectric displacement, diameter		μ	permeability
			ν	running index
E	electric field strength		ρ	specific resistance
F	force, formative area		σ	surface charge density
$G(p)$	transfer function		τ	transit time
I	current (fixed value)		φ	electric potential
$\bar{I}$	current (arithmetic mean value)		ω	angular frequency
$\hat{I}$	current (peak value)		Φ	magnetic flux
K	constant			

1 Fundamental Principles of High-Voltage Test Techniques

1.1 Generation and Measurement of High Alternating Voltages

High alternating voltages are required in laboratories for experiments and a.c. tests as well as for most of the circuits for the generation of high direct and impulse voltages. Test transformers generally used for this purpose have considerably lower power rating and frequently much larger transformation ratios than power transformers. The high voltage winding is so designed that it can withstand the routine breakdowns which generally occur on the specimen. The primary current is usually supplied by regulating transformers fed from the mains supply or, in special cases, by synchronous generators.

Most tests and experiments with high alternating voltages require precise knowledge of the value of the voltage. This demand can normally only be fulfilled by measurements on the high-voltage side of the supply.

1.1.1 Characteristic Parameters of High Alternating Voltages

The shape of $u(t)$ for high alternating voltages will often deviate considerably from the sinusoidal. In high-voltage engineering, the peak value $\hat{U}$ and the effective or root-mean-square (r.m.s.) value

$$U_{rms} = \sqrt{\frac{1}{T} \int_0^T u(t)^2 dt}$$

are of particular importance.

For high-voltage tests the quantity $\hat{U}/\sqrt{2}$ is defined as the test voltage (VDE 0432-2; IEC-Publ. 60-1)[1]. The deviations of the waveform of the high-voltage from a sine curve must satisfy the condition $\hat{U}/U_{rms} = \sqrt{2} \pm 5\%$.

Generation of High Alternating Voltages

1.1.2   Test Transformer Circuits

Transformers for generating high alternating test voltages usually have one end of the high-voltage winding earthed. Fully isolated windings are required only for special applications (e.g. symmetrical d.c. cascade).

Fig. 1.1 shows the two basic circuits for test transformers. The length of the voltage arrows indicates the magnitude of the stress on the insulation between the high-voltage winding H and the excitor winding E or the iron core F. The fully isolated winding may be earthed if necessary at either of the two terminals or at the centre tap, as shown; in the latter case, the output voltage will be symmetrical with respect to earth.

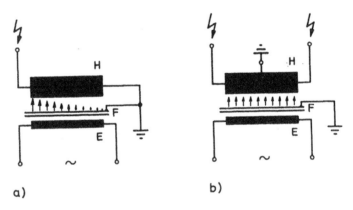

a)   b)

Fig. 1.1 Circuits of single stage test transformers
a) single pole isolated , b) fully isolated
E: Excitor winding
H: High voltage winding
F: Iron core

---

[1] VDE: Verband Deutscher Elektrotechniker. IEC: International Electrotechnical Commission.

To generate voltages above a few hundred kV, single-stage transformers according to Fig. 1.1 are now rarely used; for economical and technical reasons one employs instead a series connection of the high-voltage windings of several transformers. In such a cascade arrangement, the individual transformers must be installed insulated for voltages corresponding to those of the lower stages. Accordingly, the excitor windings of some of the transformers will have to operate at high potential.

A frequently used circuit, introduced in 1915 by *W. Petersen, F. Dessauer* and *E. Welter,* is shown in Fig. 1.2. The excitor windings E of the upper stages are supplied from the coupler windings K of the stages immediately below. The individual stages, except the uppermost, must consist of three-winding transformers. When the temperature rise [*Grabner* 1967], the curve shape [*Matthes* 1959, *Müller* 1961] and the short-circuit voltage [*Hylten-Cavallius* 1986] are determined, it should be noted that the coupling and excitor windings of the lower stages have to transmit higher powers than those of the upper ones and accordingly have to be designed for higher loading. The magnitude of the power carried by the individual windings is indicated in Fig. 1.2 in terms of multiples of P.

The calculation of the total short-circuit impedance of a cascade arrangement from data for the individual stages will be demonstrated in Appendix 4.2. Test transformers in cascade connection have already been fabricated for voltages up to 3 MV [*Frank et al.* 1991].

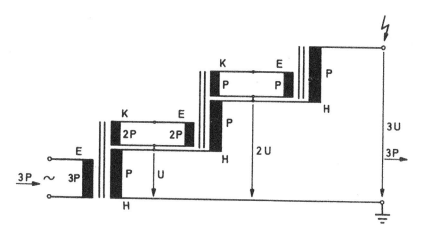

Fig.1.2 Test transformer in 3-stage cascade connection
E: Excitor winding,   H: High-voltage winding,   K: Coupler winding

1.1.3   Construction of Test Transformers[2]

For ratings of a few kVA, inductive voltage transformers can be used to generate high a.c. voltages. Low power transformers are also similar in construction to voltage transformers with the same test voltage. For voltages up to about 100 kV epoxy resin insulation is widely used; oil-impregnated paper or oil with insulating barriers and spacers are used at higher voltages. At higher power ratings, cooling of the windings becomes important, and the construction features resemble those of power transformers. Oil with barriers and oil-impregnated paper predominate as insulation. For testing of SF_6-insulated components or set-ups, transformers of totally enclosed type are preferred. In these, SF_6-impregnated foils are introduced as insulating material between the layers of the winding [*Moeller* 1975].

Test transformers with cast resin insulation have at least their high-voltage winding moulded in epoxy resin. Fig. 1.3 shows a much simplified cross-section of such a transformer.

There are numerous designs for oil-insulated test transformers. In the tank type construction, shown in Fig. 1.4a, the active parts (core and windings) are enclosed in a metal container the surface of which provides useful self-cooling. However, at high working voltages the space requirement and high cost of the bushing is a disadvantage. In the insulated enclosure transformer type, as shown in Fig. 1.4b, the active parts are surrounded by an insulating cylinder. In general, transformers of this kind contain a relatively large quantity of oil and so have large thermal time constant in the case of overloading. Heat dissipation through the insulated enclosure is very small; consequently, closed-circuit cooling by means of external heat exchangers is necessary at high continuous rating. The advantage is that no bushings are required and high-voltage electrodes with large radii of curvature can easily be fitted.

An advantageous and therefore frequently used arrangement of the active parts is shown in Fig. 1.5. It can be considered as a 2-stage cascade in which both stages have a common iron core F, which is at mid-potential and thus normally requires insulated mounting. For the symmetrical arrangement of the windings shown, E_1 or E_2 can be chosen for the primary excitation. If a cascade circuit is to be set up with a further transformer unit, via K_1, K_2, a symmetrical high voltage with respect to earth is obtained.

---

[2] Compare also *Sirotinski* 1956; *Lesch* 1959; *Potthoff, Widmann* 1965; *Prinz* 1965; *Grabner* 1967; *Beyer et al.* 1986

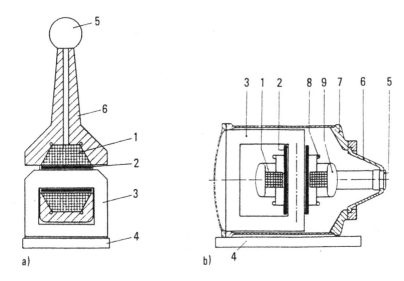

Fig. 1. 3 Cross-sectional view of test transformers
a) with cast resin insulation . b) with SF$_6$-foil insulation
1 High-voltage winding
2 Low-voltage winding
3 Iron core
4 Base
5 High-voltage terminal
6 Insulation
7 Metal housing
8 Intermediate electrode of high-voltage winding
9 High-voltage electrode of high-voltage winding

The example of Fig. 1.5 shows the voltages to earth which occur when the right high-voltage terminal is earthed.

The described arrangement is especially advantageous at very high voltages and can be set up according to the tank type design with two bushings, and also according to the insulated enclosure type design. In the latter case however, the arrangement would be turned by 90° so that the two stages lie above each other.

1.1.4   Performance of Test Transformers

The working performance of test transformers can be described with the aid of the simplified transformer equivalent circuit (Fig.1.6a). The self-capacitance C_T of the high-voltage windings and the capacitance of the connected test object (eventually inclusive of the measuring set-up), which

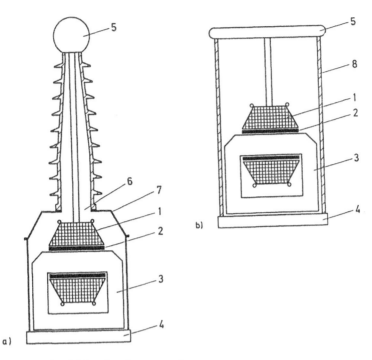

Fig. 1.4 Oil-insulated test transformers
a) tank type of construction ,   b) insulated enclosure type of construction
1 to 5 see Fig. 1.3 ,
6 Bushing
7 Metal housing
8 Insulated enclosure

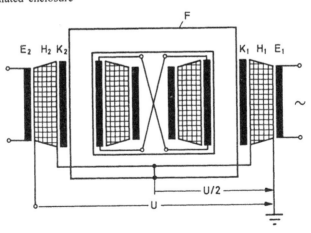

Fig. 1.5 A 2-stage cascade with common iron core at mid-potential
E_1, E_2 : Excitor windings. H_1, H_2 : High-voltage windings
K_1, K_2 : Coupler windings. F : Iron core

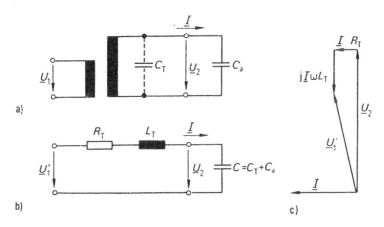

Fig. 1.6 Working performance of test transformers
a) circuit diagram, b) equivalent circuit, c) phasor diagram

represents a predominantly capacitive external load C_a, make up the load across the transformer. On the other hand, the magnetization current can be neglected as long as there is no saturation of the iron core.

Fig. 1.6b shows the equivalent circuit with the short-circuit impedance $R_T + j\omega L_T$ and the total capacitance $C = C_T + C_a$ on the high-voltage side. U_1' is the voltage due to transformation of the primary voltage U_1 to the secondary side. This equivalent circuit can also represent test transformers in cascade connection.

Since as a rule $R_T \ll \omega L_T$ and the secondary voltage U_2 is then almost in phase with the primary voltage U_1, we have:

$$U_2 \approx U_1' \frac{1}{1 - \omega^2 L_T C}.$$

The expression $1/(1 - \omega^2 L_T C)$ is always > 1. Thus series resonance leads to a capacitive enhancement of the secondary voltage. The amount of capacitive voltage enhancement can easily be calculated from the transformed short-circuit voltage u_k of the transformer, for the case when the capacitive load C just takes rated current I_n at rated voltage U_n and nominal frequency, as:

$$u_k = \frac{I_n \omega L_T}{U_n} = \omega^2 L_T C .$$

Thus a test transformer with $u_k = 20\,\%$ will show a voltage enhancement of 25% at nominal frequency when a capacitive load takes the rated current.

This voltage enhancement has to be taken into account, particularly for test transformers with high values of transformed short-circuit voltage, and above all when used at higher frequencies. There is then no longer a fixed ratio of primary to secondary voltage; for this reason determination of the high-voltage output by voltage measurement on the primary side of the transformer is inadmissible. This measurement would indicate values well below the real ones, and the object as well as test transformer could be endangered.

Test transformers, especially in cascade connection, represent spatially extended networks capable of oscillation. Harmonics of the primary voltage and the magnetization current may excite natural oscillations at various frequencies, and this can lead to considerable distortion of the secondary voltage. These distortions from the ideal sine curve due to the harmonics of the high-voltage depend upon the load current and the value of the set voltage.

1.1.5   Compensation for a Capacitive Load

With large capacitive loads, e.g. while testing high voltage cables, it is advantageous to compensate for the reactive kVA of the specimen. Fig. 1.7 shows the possible circuits for this. Compensation with additional switchable coils on the primary side (1) is common due to economic reasons. The connected power rating from the supply network and the regulating transformer can be dimensioned for a lower power rating, whereas the high voltage transformer must transmit the full power. Compensating coils on the high voltage side (2) are often an uneconomical solution though the test transformer can be designed for a lower rating since the coils must be

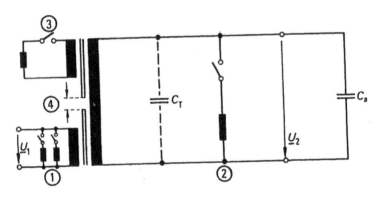

Fig. 1.7 Compensation of the capacitive reactive kVA with switchable coils (1) on the primary side, (2) on the high-voltage side, (3) through a tertiary winding with coils built- in into the tank or (4) with adjustable air-gap in the transformer core

dimensioned similar to the high voltage winding of the test transformer. Solution (3) has found acceptance in multi-stage cascades with very large power ratings.

Compensation of the capacitive reactive kVA with the help of an adjustable air-gap in the iron core (4) has been resorted to in single stage set-ups of the tank type construction.

1.1.6 Generation of High Voltage in a Series Resonant Circuit

Besides this conventional method, compensation of the reactive power requirement of the specimen in a series circuit is also possible with parallel connected inductive coils. A variable high voltage inductor together with an additional transformer of comparatively lower secondary voltage forms a series resonant set-up (Fig. 1.8).

The variable inductor is made possible by an adjustable air-gap in the iron path. An advantage of the series resonant circuit is that it delivers a high voltage with low distortion, has a low short circuit power rating, and due to the working principle of the resonant circuit, achieves an almost complete compensation of the reactive power requirement of the specimen. Thus it is possible in the series resonant circuit to design the connected power requirement as well as that of the regulating transformer correspondingly low. While setting it up, it

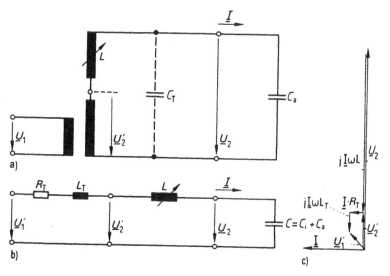

Fig 1.8 Working principle of series resonant setups
a) circuit diagram, b) equivalent circuit c) phasor diagram

is necessary to pay attention especially to the stray flux lines in the air-gap and the vibrational noise.

For high voltages (above 500 kV) the various inductor is realized by a large number of inductors in insulated-tank type construction stapled one above the other. For e.g., series resonant circuits upto 2.2 MV and 10MVA have been built with 6 inductors in series.

As may be seen clearly from the phasor diagram of the series resonant setup (Fig 1.8c), the overvoltage caused by the inductance L in a capacitively loaded test transformer is intentionally made use of for obtaining a high voltage on its secondary side. In practice, a quality factor of 50...100 is achievable for the resonant circuit. Such setups are, due to constraints arising out of their working principle, not suitable for pure resistive loading e.g., as encountered in pollution testing. The heavy pulse like discharges result in voltage dips.

For *in-situ* testing of cables or SF$_6$ setups, a series resonant set-up with a fixed (non-variable) inductance and variable supply frequency has been realized. A disadvantage of this type is that the test is not conducted at a rated frequency. On the contrary, an advantage of the set-up for *in-situ* testing is the comparatively small and light inductance that, depending on the test voltage level, can be arranged in series or parallel [*Zaengl et al.* 1982].

The Tesla-transformer [*Marx* 1952; *Heise* 1964], named after its inventor also belongs to the resonant circuits. The circuit consists of a primary and a secondary oscillatory circuit, which are loosely coupled with one another magnetically. Such a set-up capable of oscillating is excited by periodic discharge of the primary side capacitor via a spark gap to oscillate at high frequencies. Based on the chosen circuit parameters and the transformation ratio between the primary and secondary windings, voltages upto 1 MV and above can be generated with Tesla transformers. These have been made use of in breakdown testing of porcelain pin-type insulators.

1.1.7   Requirements on a Voltage Source for Pollution Tests

Currents upto a few amperes flow through the specimen during pollution testing [*Rizk, Bourdage* 1985]. The requirement valid for a meaningful test is that the pollution layer current during a withstand test does not affect the test voltage to an impermissible extent. Fig 1.9 shows the equivalent circuit and the basic performance of the test - circuit during pollution testing.

The pollution layer current i_H results in a voltage drop Δu_2 across the internal resistance of the voltage source. In order that this voltage drop, referred to the open circuit voltage u_{20}, does not exceed 10%, requirements are imposed on a voltage source for pollution investi-gation, especially on its short circuit current rating (IEC- Publ. 507). The short circuit current $I_{k\,min}$ shall be 11 times greater than the highest magnitude of the pollution layer current $I_{H\,max}$ during a withstand test. This requires short circuit currents of the order of 5.5 to 22 A for pollution layer currents of 0.5 to 2 A. Further requirements placed on voltage sources for pollution tests corresponding to IEC-Publ. 507 are usually met [*Köhler* 1988].

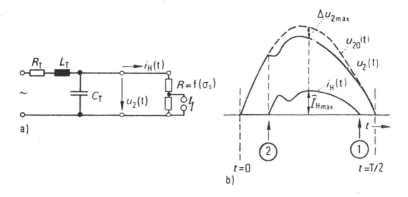

Fig 1.9 Pollution test (σ_s - layer conductance of pollution layer)
a) equivalent circuit of test set-up and the specimen,
b) basic wave form of test voltage u_2 and pollution layer current i_H
((1) quenching, (2) restriking of partial arcs)

1.1.8 Measuring Technique for Determining the Characteristic Parameters of a Test Setup

The characteristic parameters of the equivalent circuit of a test set up, including that of the regulating transformer, can be determined in the actual setup with high voltage (Fig.1.10).

While switching-in the energising voltage with a switch S, the test voltage u_2 oscillates (resonates) with a natural frequency f_e of the test voltage source, wherein

$$f_e = \frac{1}{2\pi}\sqrt{\frac{1}{L_T.C_T} - \left(\frac{R_T}{2.L_T}\right)^2}.$$

Usually, the term $(R_T/2.L_T)^2$ can be neglected, such that:

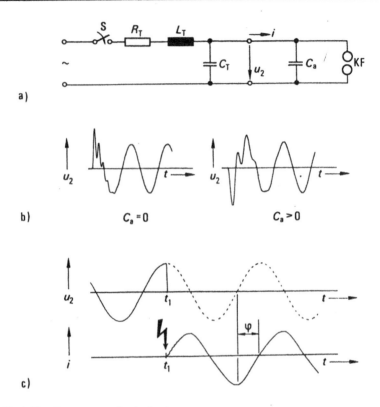

Fig.1.10 Parameters of a test setup
a) equivalent circuit,
b) $u_2(t)$ without and with C_a,
c) u_2 and i from short circuit measurement ($C_a=0$, t_1 - breakdown of KF)

$$f_e = \frac{1}{2\pi} \cdot \frac{1}{\sqrt{L_T \cdot C_T}}$$

A second attempt at switching-in with a connected known capacitor C_a results in:

$$f_e' = \frac{1}{2\pi} \cdot \frac{1}{\sqrt{L_T(C_T + C_a)}} \cdot$$

From the measured natural frequencies the inductance L_T and the parallel capacitance C_T can be calculated to be:

$$C_T = \frac{1}{(f_e/f_e')^2 - 1} \cdot C_a \quad ; \quad L_T = \frac{1}{C_T} \left(\frac{1}{2\pi \cdot f_e}\right)^2 \cdot$$

In order to determine the value of the ohmic internal resistance R_T, a special short circuit measurement is conducted with the sphere-gap (KF). The voltage before breakdown and current immediately thereafter are measured. Fig.1.10c shows an oscillogram of short circuit measurement with extrapolation of the voltage waveform , from which the phase angle φ is to be determined. We have:

$$R_T = \frac{\omega L_T}{\tan \varphi}.$$

1.1.9   Protection of Test Transformers

Test transformers are endangered by steep voltage collapses on the specimen(non linear voltage distribution, internal resonance) as well as over voltages(flash over of a specimen with rapid recovery of the arc insulation strength). Damping resistors (a few 100 Ω) on the high-voltage side are used as protection against steep voltage collapse on the specimen. These are especially important in the case of test transformers directly coupled to SF_6 - setups. Fast short circuiting with a thyristor switch on the primary side of the high voltage side of the transformer and simultaneous switching - off have proved to be successful as protection against overvoltages. Triggering of the short circuit switch can be effected while exceeding a certain voltage or at the appearance of a short circuit current within a few ms [Hylten - Cavallius 1986].

Measurement of High Alternating Voltages[3]

1.1.10   Peak Value Measurement with Sphere-Gaps

Breakdown of a spark gap occurs within a few μs once the applied voltage exceeds the " static breakdown discharge voltage". Over such a short period the peak value of a power frequency voltage can be considered to be constant. Breakdown in gases will therefore always occur on the peak of low frequency a.c. voltages. With approximately homogeneous field gaps, for which the breakdown discharge times are particularly short, this behaviour is followed quite well to higher frequencies. Consequently the peak values of high a.c. voltages of frequencies up to about 500 kHz can

---

[3] Comprehensive treatment, among others in *Craggs. Meek* 1954; *Sirotinski* 1956; *Potthoff, Widmann* 1965; *Schwab* 1969; *Kuffel, Zaengl* 1984; *Beyer et al.* 1986.

be determined from the gap spacing at breakdown of measuring spark gaps in atmospheric air.

Fig. 1.11 shows the two basic arrangements of sphere-gaps for measuring purposes. The horizontal arrangement is usually preferred for sphere diameters $D < 50$ cm used for the lower voltage ranges; with the larger spheres the vertical arrangement is chosen; it is most suitable for measuring voltages with reference to earth potential only.

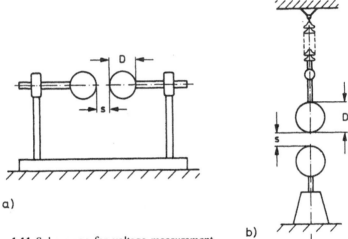

Fig. 1.11 Sphere-gap for voltage measurement
a) horizontal, b) vertical arrangement

The published specifications (VDE 0433-2; IEC-Publ. 52) prescribe minimum clearances from objects disturbing the electric field and tabulate breakdown voltages for standard conditions and various sphere diameters D as a function of the gap spacing s:

$$\hat{U}_{d0} = f(D,\ s)$$

The values are valid for an atmospheric pressure of $b = 101,3$ kPa and a temperature $\vartheta = 20°C$. Humidity has no significant influence on the breakdown voltage of sphere gaps. Fig. 1.12 demonstrates the dependence of breakdown voltage upon gap spacing for various sphere diameters. For measurement with sphere gaps, with increasing ratio s/D the field becomes increasingly inhomogeneous; at the same time the influence of the gap surroundings becomes greater, and so does the scatter in the values of breakdown voltages. Evidently the ratio s/D may not be too large. The minimum sphere diameter D for measurement of a voltage of amplitude $\hat{U}$ can be estimated from the following relationship:

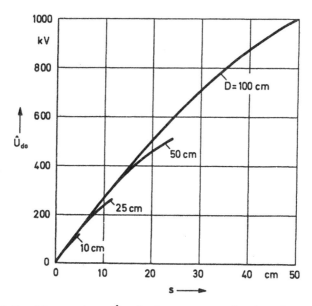

Fig. 1.12 Breakdown voltage $\hat{U}_{do}$ of sphere-gaps as a function of gap spacing s, for various sphere diameters D

$D \geq \hat{U}$ with D in mm and $\hat{U}$ in kV.

It should be pointed out that during these measurements the tabulated values are only valid as long as the minimum clearances between the gap and the other parts of the setup are maintained.

Since the breakdown voltage $\hat{U}_d$ is proportional to the relative air density d in the range 0.9...1.1, the actual breakdown voltage $\hat{U}_d$ at air density d may be found from the tabulated value $\hat{U}_{do}$ by applying the following formula:

$$\hat{U}_d \approx d.U_{d0} = \frac{b}{101.3} \cdot \frac{273+20}{273+\vartheta} \hat{U}_{d0} = 2.89 \frac{b}{273+\vartheta} \hat{U}_{d0}$$

with b and ϑ in kPa[4] and °C respectively.

Even under apparently ideal conditions, having made allowance for such factors as the air density, minimum clearances, smooth exactly spherical electrode surface and proper adjustment of the spacing, a measuring uncertainty of 3 % remains. Sphere-gaps are now rarely used for measuring voltages above 1 MV, because they require excessive space and are

---

[4] 1 kPa = 10 mbar ≈ 7.5 Torr

expensive. Continuous voltage measurement is obviously impossible with sphere-gaps, since the voltage source is short-circuited at the instant of measurement. The method is suitable especially for control of complete measuring arrangements with high-voltage.

In spite of their disadvantages, sphere-gaps can be useful and versatile devices in a high-voltage laboratory. Apart from voltage measurement, they can also be used as voltage limiters, as voltage-dependent switches, as pulse sharpening gaps and as variable high-voltage capacitors, etc.

1.1.11 Voltage Measurement Using High-Voltage Capacitors

High-voltage capacitors are specially suited for reduction of high alternating voltages to values measurable with instruments. In order to keep the loading on the voltage source as low as possible, the high voltage capacitor C_1 should be as small as possible. The accuracy of voltage measurement with high voltage capacitors is then limited by the surrounding that can affect the capacitor C_1, represented in the equivalent circuit by the earth capacitance C_E. Taking the earth capacitance for a high voltage capacitor into account results in the equivalent circuit of Fig. 1.13.

The measuring circuit is connected at the low-voltage output terminal the high-voltage capacitor could be the capacitance of a high-voltage divider (1.1.13) or the series impedance of an ammeter (1.1.12). For the current flowing through the measuring circuit, determined by the primary capacitance C_1, the earth capacitance C_E reduces to the effective primary capacitance C to;

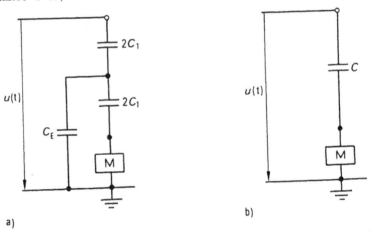

a)   b)

Fig. 1.13 Equivalent circuits of a high voltage capacitor
a) with earth capacitance C_E, b) with equivalent capacitance C

$$C = C_1 \frac{1}{1 + \dfrac{C_E}{4 \cdot C_1}} \approx C_1 \left(1 - \frac{C_E}{4 \cdot C_1} \right)$$

Under the assumption of homogeneous distribution of earth capaci-tance it can be shown that C_E is equal to 2/3 of the total earth capacitance C_e acting at C_1. For vertical cylindrical dividers, C_e can be calculated at a value of 12...20 pF/m height.

The effect of change of capacitance must remain small to ensure adequate measuring accuracy. This can be achieved in practice by making the high voltage capacitors static(always the same C_e) or with low earth capacitance(screened) or with sufficiently large C_1. For a design with $C_1 > 2C_e$ in the usual set-ups, the measuring discrepancy can be calculated to be less than 1% [*Lührmann* 1970].

1.1.12 Peak Value Measurement with a High-Voltage Capacitor as Series Impedance

The circuit shown in Fig. 1.14, suggested by *Chubb* and *Fortescue* in 1913, is well suited for exact and continuous measurement of the peak value of high a.c. voltage against earth.

A charging current i, given by the rate of change of the applied voltage $u(t)$ to be measured, flows through the high-voltage capacitor C and is

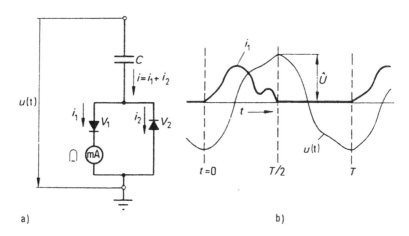

a)   b)

Fig. 1.14 Peak voltage measurement according to *Chubb* and *Fortescue*
a) circuit , b) current and voltage curves

passed through two antiparallel rectifiers V_1 and V_2 to earth. The arithmetic mean value $\bar{I}_1$ of the current i_1 in the left-hand branch is measured with a moving-coil instrument; as shown below, provided that certain conditions are fulfilled, this current is proportional to the peak value $\hat{U}$ of the high-voltage.

If the behaviour of the rectifiers is assumed ideal, then for the conducting period of V_1 we have:

$$i_1 = i = C\frac{du}{dt} \qquad for\ t = 0...T/2$$

$$\bar{I}_1 = \frac{1}{T}\int_0^T i_1 dt = \frac{1}{T}\int_{u(0)}^{u(T/2)} C\,du = \frac{C}{T}[u(t = T/2) - u(t = 0)].$$

If the voltage is symmetrical with reference to the zero line :

$$u(t{=}T/2) - u(t{=}0) = 2\hat{U}$$

and with $T = 1/f$, we obtain

$$\hat{U} = \bar{I}_1\frac{1}{2fC}.$$

If a circuit with full-period rectification (Graetz circuit) is used instead of the half-period rectifier circuit shown in the figure, the factor 2 in the denominator of the above equation should be replaced by 4.

For the derivation of this expression, it was not assumed that $u(t)$ is a sinusoid, though when passive rectifiers (especially semiconductor diodes) are used, we have to demand that the high voltage to be measured does not have more than one maximum per half-period. The use of synchronous mechanical rectifiers or controllable rectifiers (oscillating contacts, rotating rectifiers) allows correct measurement of alternating voltages with more than one maximum half-period.

Oscillographic monitoring of the high-voltage shape is necessary and is usually done by observing the current i_1, which may have one crossover only in each half-period. As the frequency f, the effective high voltage capacitance C and the current $\bar{I}_1$ can be determined precisely, measurement of symmetrical a.c. voltages using technique of *Chubb* and *Fortescue* with the appropriate outlay is very accurate, and is suitable for the calibration of other peak voltage measuring devices [*Boeck* 1963]. A disadvantage for technical routine measurements is the dependence of the reading upon the frequency and the need to monitor the curve.

Based on the above principle, measuring set-ups with electronic compensators / sensors have been developed which, by means of switching principles, have further reduced the error influences or eliminated them altogether. With such measuring circuits, both positive and negative peak values of asymmetrical alternating voltages and also the frequency can be measured. The uncertainty during alternating voltage measurement can thereby be reduced to very low values [*Marx, Zirpel* 1990].

1.1.13 Peak Value Measurement with Capacitive Voltage Dividers

Several rectifier circuits have been developed which, together with the extension of the high voltage capacitor by means of a low voltage capacitor so as to form a capacitive voltage divider permit the measurement of peak values of high alternating voltages. Compared with the circuit of *Chubb* and *Fortescue*, most of these methods have the advantage that the reading is practically independent of frequency, and multiple extremes per half-period of the voltage to be measured can be permitted.

The half-wave circuit shown in Fig. 1.15 is practically simple and also sufficiently accurate for most purposes. In this circuit, the measuring capacitor C_m is charged to the peak value U_2 of the lower arm voltage $u_2(t)$ of a capacitive divider. The resistor R_m which discharges the capacitor C_m is necessary to ensure an adequate response to reductions in the applied voltage. The choice of time constant for this discharge process is determined by the desired response of the measuring arrangements, whereby the internal resistance of the connected measuring instrument must be taken into account. In general, one chooses:

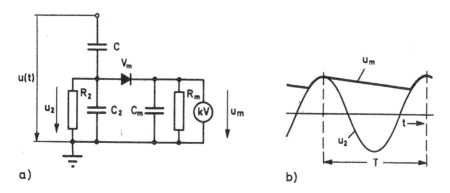

Fig. 1.15 Peak voltage measurement with capacitive divider
a) circuit. b) general form of the voltage

$R_m \, C_m < 1s.$

On the other hand, this time constant must be large compared with the period $T = 1/f$ of the alternating voltage to be measured, since otherwise the voltage u_m due to the discharge of C_m is not sufficiently constant as is indicated in Fig.1.15b. The appropriate condition here is:

$R_m \, C_m \gg 1/f.$

The resistance R_2 parallel to C_2 is necessary in order to prevent charging of C_2 by the current flowing through the rectifier V_m. The value of R_2 must be chosen in such a way that the direct voltage drop across R_2 which causes d.c. charging of C_2 remains as small as possible, thus we must have:

$R_2 \ll R_m \, .$

On the other hand, the capacitive divider ratio should not be affected much by R_2, so that :

$$R_2 \gg \frac{1}{\omega C_2}$$

Provided all these conditions can be satisfied, the relation between the peak value of the high voltage and the indicated voltage U_m is given by:

$$\hat{U} = \frac{C + C_2}{C} \hat{U}_m.$$

The indicating instrument should have a high input impedance; electrostatic voltmeters, high sensitivity moving-coil instruments and resistance or electrometer amplifiers with analog or digital indication are suitable. Measuring range changes are usually effected by changing C_2. Digital measuring instruments with built-in microprocessors enable the determination of the real effective(r.m.s) value and also a Fourier analysis of the waveform. For the A/D converter, a vertical resolution of 12 bit at a sampling frequency of about 10 kHz is adequate for measuring alternating voltages up to 200 Hz.

The postulates made above for the relative values of the circuit components are not quite compatible and limit the obtainable accuracy, particularly at low frequencies. The properties can be improved with more elaborate circuitry [Zaengl, Völcker 1961].The overall achievable accuracy, however, not only depends upon the properties of the low-voltage measuring circuit, but also upon those of the high-voltage capacitor.

1.1.14 Measurement of R.M.S. Values by Means of Electrostatic Voltmeters

When a voltage $u(t)$ is applied to an electrode arrangement, such as the one shown in Fig. 1.16a for example, the electric field produces a force $F(t)$ which tends to reduce the spacing s of the electrodes. This attractive force can be calculated from the change of energy of the electric field. The capacitance C of the arrangement is thereby dependent on the spacing s. The time variant form of the force is obtained from the law of conservation of energy

$$dW + F \, ds = 0$$

assuming disconnection of the voltage source [*Küpfmüller* 1965]. Taking into account that the charge $C.u(t)$ is independent of s, it follows:

$$F(t) = -\frac{dW(t)}{ds} = \frac{1}{2}u(t)^2 \frac{dC}{ds}.$$

If the arithmetic mean value $\bar{F}$ of the force is calculated from this expression, the linear relationship between $\bar{F}$ and the square of the r.m.s. value of the applied voltage is apparent:

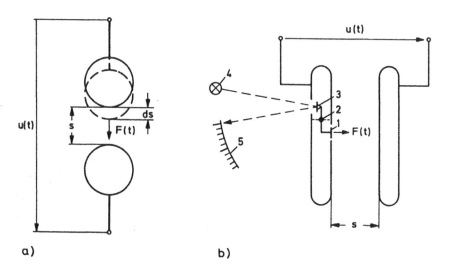

a)   b)

Fig. 1.16 Electrostatic voltmeters for high voltages
a) using spherical electrodes (after Hueter)
b) using a movable electrode segment (after Starke and Schröder)
1: Movable electrode, 2: Axis, 3: Mirror, 4: Light source, 5: Scale

$$\overline{F} = \frac{1}{2} \cdot \frac{dC}{ds} \cdot \frac{1}{T} \int_0^T u(t)^2 \, dt \sim U^2_{eff} \, .$$

The influence of the factor dC/ds depends upon the way in which the force $\overline{F}$ is translated into an indication. In general, dC/ds changes over the measuring range, so that the deflection no longer shows strict quadratic dependence.

As an example of an electrostatic measuring device, the design of *Starke* and *Schröder* is shown in a simplified form in Fig. 1.16b. The force $F(t)$ acts on a small plate 1 mounted on a cranked lever with a central pivot; at the other end of the lever is a small mirror which deflects a light beam for the optical indication. The taut band suspension 2 provides the restoring torque.

Electrostatic voltmeters are characterized by their very high internal resistance and very small capacitance; they are thus also useful for the direct measurement of high-frequency high-voltages extended to the MHz region. However, they are seldom used since the requirement of space as well as cost are extremely high at high-voltages.

1.1.15 Measurement with Voltage Transformers

High alternating voltages can be measured extremely accurately with voltage transformers. Although these devices are widely used in power supply networks, they are rarely used in laboratories for measurements of voltages above 100 kV.

The basic circuits of single pole isolated inductive and capacitive voltage transformers for the measurement of voltages with respect to earth, are shown in Fig. 1.17 together with terminal markings as per appropriate specifications (VDE 0414-2).

Inductive voltage transformers for very high voltages can be built only at great expense since, for the comparatively low test frequency of 50 Hz, the product of magnetic flux and number of turns of the high-voltage winding, by the laws of induction, takes very large values. This leads to expensive designs.

The type of capacitive voltage transformer used extensively in supply networks is often considered unsuitable for normal testing work, mainly because it imposes a high capacitive load upon the voltage source.

Inductive and capacitive voltage transformers are therefore utilised in laboratories especially for calibration purposes. The secondary voltage of a voltage transformer will reproduce the shape of the primary voltage,

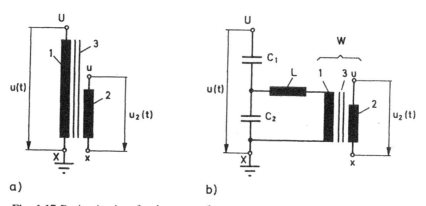

Fig. 1.17 Basic circuits of voltage transformers
a) inductive voltage transformers, b) capacitive voltage transformers
1: Primary winding, 2: Secondary winding, 3: Iron core, C_1, C_2 : Divider capacitors, L: Resonance inductor, W: Matching transformer (markings as under a))

irrespective of the secondary load. Depending upon the type of measuring device connected, it is possible to measure the peak value, the r.m.s. value or the high-voltage curve.

1.2 Generation and Measurement of High Direct Voltages

There are numerous applications for high direct voltages in the laboratory, such as for the testing of HVDC transmission equipment, for the investigation and testing of insulating arrangements with high capacitance, e,g, capacitors or cables, and fundamental investigations in discharge physics and dielectric behaviour. Technical uses include the generation of X-rays, precipitators, paint spraying and powder coating. The most common generation methods of high direct voltages employ rectification of high alternating voltages, often using voltage multiplication; electrostatic generators are also in use. The high direct voltages are usually measured by means of high resistance measuring resistors or by electrostatic voltmeters.

1.2.1   Characteristic Parameters of High Direct Voltages

The d.c. test voltage is defined as the arithmetic mean value (VDE 0432-2; IEC Publ. 60-1):

$$U = \frac{1}{T} \int_0^T u(t)\mathrm{d}t.$$

Periodic fluctuation of the direct voltage between the peak value $\hat{U}$ and the minimum value U_{min} are given in terms of the ripple amplitude:

$$\delta U = \frac{1}{2}\left(\hat{U} - U_{min}\right).$$

The expression $\dfrac{\delta U}{U}$ is called the "ripple factor".

For high-voltage tests, the ripple factor shall be lower than 3% (VDE 0432; IEC Publ. 60-1).

Generation of High Direct Voltages[5]

1.2.2   Properties of High-Voltage Rectifiers

As elements for the rectification of high alternating voltages, usually series-connected stacks of semiconductor diodes or high vacuum valves are made use of (Fig. 1.18).

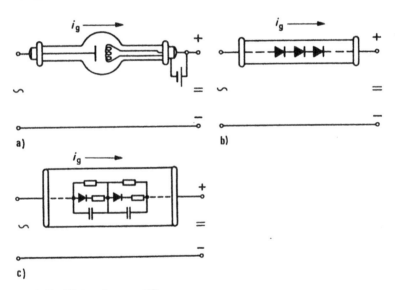

Fig. 1.18 High-voltage rectifier
a) high-vacuum valve
b) semiconductor diode
c) semiconductor diode with protective circuitry

---

[5] Comprehensive treatment among others in *Craggs, Meek* 1954; *Sirotinski* 1956; *Lesch* 1959; *Kuffel, Zaengl* 1984; *Beyer et al.* 1986

In high-vacuum rectifiers, the current is carried by electrons emitted from a thermionic cathode and accelerated towards the anode by the electric field. These rectifiers are available for use with peak inverse voltages of up to 100 kV. Although in laboratory practice high-vacuum rectifiers have been replaced by semiconductor rectifiers, which are more convenient to use as no provision need be made for cathode heating, the former are of great advantage in X-ray installations since they can function as X-ray tubes at the same time.

In contrast to high-vacuum rectifiers, semiconductor diodes are not true valves since they allow a small but finite current flow in the blocked condition. The following guiding values may be given for inverse voltages and forward currents of the more commonly used semiconductor rectifiers:

Semiconductor material	Selenium	Germanium	Silicon
Peak inverse voltage per element	30-50	150-300	1000-2000 V
Loading capacity of the depletion layer	0.1-0.5	50-150	50-150 A/cm^2

For high-voltage applications today, mostly Si-diodes with appropriate protective circuitry (Fig. 1.18c) are made use of.

1.2.3   The Half-Wave Rectifier Circuit

The simplest circuit for the generation of a high direct voltage is the half-wave rectification shown in Fig. 1.19. A load R is supplied from a high-voltage transformer T via a rectifier V. We assume that the secondary voltage u_T of the transformer is a sinusoid and the rectifier is ideal, i.e. with zero forward resistance and zero reverse current. Depending on whether the smoothing capacitor C shown by a dashed line, is connected or not, curves of Figs. 1.19b and 1.19c show the voltage across the load for steady-state conditions.

The circuit without the smoothing capacitor C will give a pulsating direct voltage with the following characteristic values:

$$\hat{U} = \hat{U}_T \, ; \overline{U} = \frac{1}{\pi} \hat{U} \, ; U_{rms} = \frac{1}{2\sqrt{2}} \hat{U}.$$

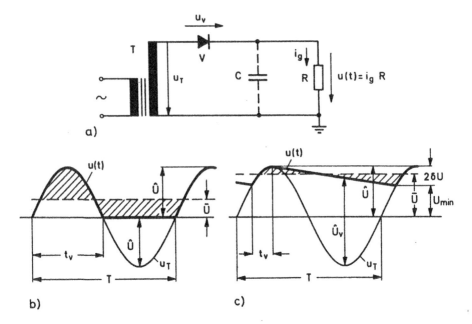

Fig. 1.19 Half-wave rectification with ideal circuit elements
a) circuit diagram,
b) output voltage curve without smoothing capacitor C,
c) output voltage curve with smoothing capacitor C

The conducting period t_v of the rectifier is equal to half the period T of the alternating voltage. The peak inverse voltage across the rectifier in the reverse direction is:

$$\hat{U}_v = \hat{U}_T.$$

For the circuit with the smoothing capacitor C a smoother direct voltage U with ripple voltage δU is obtained. We have:

$$\hat{U}_v = \hat{U}_T; \quad U \approx \hat{U} - \delta U.$$

The better the smoothing of the voltage, the shorter the current flow period t_v will be. Thus, during the conducting period of the rectifier, only a short forward current pulse flows each time, the peak inverse voltage being:

$$\hat{U}_v \approx 2\hat{U}_T.$$

Referring to Fig. 1.19c, the ripple voltage δu can easily be calculated for

$$t_v \ll T = \frac{1}{f} \quad \text{and} \quad \delta U \ll \overline{U}$$

The exponential discharge of C during the blocking period of V can be replaced by a straight line. From the change of charge on the smoothing capacitor during the blocking period, we have:

$$2 \cdot \delta U C \approx \int_0^T i_g \, dt = T \cdot I_g$$

$$\delta U \approx I_g \frac{1}{2fC}.$$

In full-wave rectification, the time intervals between successive recharging, and thus the ripple voltage δU, are reduced to one half. The usual methods for reducing ripple voltage in δU rectifier circuits are increasing the size of the smoothing capacitor, the frequency and the number of phases. In laboratory setups frequencies up to some 1000 Hz are frequently used and ripple factors of a few percent (<5%) are common.

In the design of circuits, the non-ideal behaviour of rectifiers has to be taken into account; in particular, the forward voltage drop for current flow in the conduction direction must be allowed for. A non-linear relationship between the direct current $\bar{I}_g$ and the direct voltage U is the consequence. Fig. 1.20 shows the typical load characteristic for a semiconductor rectifier. For $\bar{I}_g = 0$, the ideal no-load voltage U_{i0} is the peak voltage of the transformer, $U_{i0} = \hat{U}_T$. Linear extrapolation of the load characteristic for high currents to $\bar{I}_g = 0$ gives an intercept with the ordinate at a value $\bar{U}_0$ which is lower than $\bar{U}_{i0}$ by the value U_1 for each element, practically independent of the current. The output voltage of a rectifier of n elements

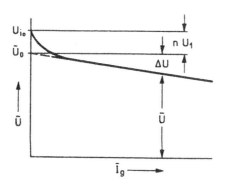

Fig. 1.20   Load characteristic for semiconductor rectifiers

in series can therefore, except for very low currents, be described by the equation:

$$U = U_0 - \Delta U = (U_{i0} - nU_1) - kI_g,$$

where k is a proportionality factor depending upon the type of rectifier; the voltage U_1 is of the order of 0.6 ... 1.2 V.

1.2.4 Voltage Multiplier Circuits

The most widely used multiplier circuits will now be described, assuming idealized elements. A common property of all the circuits considered here is that they are only able to supply relatively low currents and are therefore not suitable for high current applications such as high-voltage direct current transmission. The voltage curves are shown to illustrate the working principles of the various circuits. For simplification the excitation windings of the high-voltage transformers T have been omitted in the circuit diagrams.

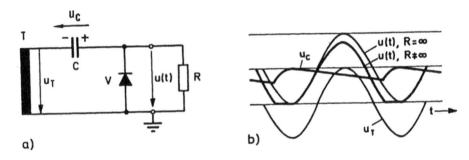

a)   b)

Fig. 1.21 Villard circuit
a) circuit diagram, b) voltage curve

Villard Circuit:

This circuit, shown in Fig. 1.21, is the simplest doubling circuit. The blocking capacitor C is charged to the peak value $\hat{U}_T$ and thus increases the potential of the high-voltage output terminal with respect to the transformer voltage by this amount. For no-load conditions, we have:

$$U = \hat{U}_T \; ; \; U = 2\hat{U}_T \; ; \; \hat{U}_v = 2\hat{U}_T.$$

Smoothing of the output voltage $u(t)$ is not possible.

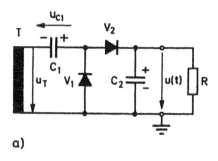 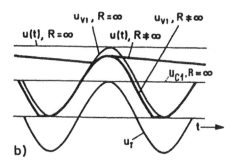

Fig 1.22 Greinacher doubler circuit
a) circuit diagram, b) voltage curve

Greinacher Doubler-Circuit:

Fig. 1.22 shows the extension of the Villard circuit by a rectifier V_2, which enables the smoothing capacitor C_2 to be connected. For no-load conditions, we have:

$$U = \hat{U} = 2\hat{U}_T; \; \hat{U}_{v1} = \hat{U}_{v2} = 2\hat{U}_T.$$

The sum of the peak inverse voltages of the rectifiers in this circuit is twice the output voltage $\bar{U}$. This is true of any rectifier circuit providing a smooth direct voltage.

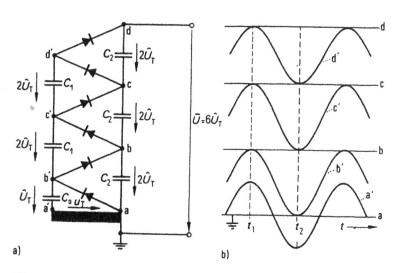

Fig 1.23 Greinacher cascade circuit (no-load condition)
a) circuit diagram, b) ideal voltage curve (without load)

Greinacher Cascade Circuit :

This circuit, suggested in 1920 by *H. Greinacher* and which is also known as the *Cockroft-Walton* multiplier, is the most important method for the generation of very high direct voltages. It is an extension of the Greinacher doubler-circuit.

Fig 1.23 shows as example a single phase three-stage circuit . The voltages indicated apply for ideal circuit elements and no-load condition. To achieve a more uniform voltage drop, it proves useful to choose the capacitance C_0 of the lowest unit of the so-called blocking capacitor stack to be twice that of the capacitors C_1 above it. The capacitor stack comprising the elements C_2 in series serves, among other functions, as a smoothing capacitor. Usually, the capacitances C_1 and C_2 are chosen to be equal in magnitude.

When the cascade unit is loaded, voltage drops occur due to the threshold voltage and the internal resistance of the diodes as also the charging / discharging mechanisms. With ideal elements and $C_0 = 2C_1 = 2C_2 = 2C$, the voltage drop ΔU due to the charging / discharging processes works out to:

$$\Delta U = \frac{\bar{I}_g}{f.C} \frac{8n^3 + 3n^2 + n}{12}$$

The voltage drop ΔU increases with the number of stages n. The ripple voltage δU can be calculated [*Beyer et al* 1986] as:

$$\delta U = \frac{\bar{I}_g}{f \cdot C} \cdot \frac{n(n+1)}{4} .$$

Fig 1.24 indicates the definitions of ΔU and δU (a) and an example of the loading range of a test rectifier (b). It may be noticed that with a given number of stages n, only a restricted loading range can be realized as per specificational requirements (e.g., $\delta U < 5\%$), due to the voltage drop ΔU and the ripple factor δU. Especially at low voltages, the loading range narrows down appreciably since the amplitude of the ripple voltage, independent of the voltage is constant for constant current. In order to achieve a low ripple voltage with multi-stage cascades, even at low voltages, it is necessary that one chooses, depending on the voltage, the minimum number of stages. Short circuiting of two stages leads to an appreciable extension of the loading range at low voltages.

Due to the strong dependence of ΔU and δU on n, it is common practice to choose as high a stage voltage as possible. A voltage of about 400 kV results in an economic optimum solution for the generation of high

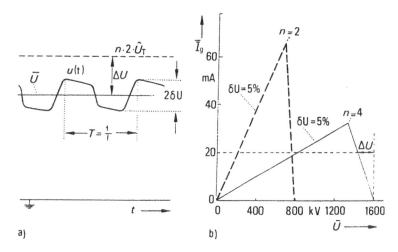

Fig 1.24 Greinacher cascade circuit under load
a) output direct voltage $u(t)$.
b) loading range of a n - stage Greinacher cascade
$(\hat{U}_T = 200$ kV, $f = 50$Hz, $C_1 = C_2 = C_0 / 2 = 60$ nF$)$

direct voltages from 1 to 2 MV. Single phase rectifier cascades up to about 2 MV have been built as test setups with a frequency of 50 or 60 Hz and currents up to a few 10 mA. For physical applications (low smoothing capacitor, low ripple voltage) or rectifier source for pollution testing (large currents), the Greinacher cascade is operated at higher frequency (2 kHz to 15 kHz) and is often built up as a multi phase unit with a smoothing capacitor stack and more number of blocking capacitor stacks.

The equivalent circuit of a 2 phase (symmetrical) Greinacher cascade with 4 stages is shown in Fig 1.25 [*Baldinger* 1959]. Appreciably favourable values of ΔU and δU are obtained in this circuit when compared with a single phase circuit. In multi-stage set-ups, sphere-gaps (or lightning arrestors) connected parallel to the topmost diodes and the energising transformers are essential for the protection of the elements. The specimen is usually connected to the Greinacher cascade through a damping resistor R. Polarity of the direct voltage can be changed by reversing the diodes. Greinacher cascades in encapsulated, SF_6- insulated type of construction have been constructed up to voltages of 5 MV and up to 2.4 MV in air.

Cascade with transformer support

For high currents larger than 100 mA it is also possible to have a series connection of individual rectifier circuits as shown in Fig 1.26. In this

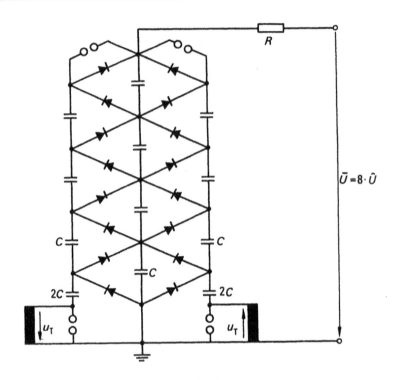

Fig 1.25   4-stage symmetrical Greinacher cascade

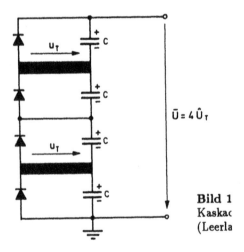

Bild 1
Kaskac
(Leerla

Fig 1.26 Cascade with transformer support (no load)

way, low ripple voltages and voltage drops can be achieved even when output currents are high. The alternating current inputs to the individual circuits must be provided at the appropriate high potential; this can be achieved either by means of isolating transformers or individual alternators driven by insulating shafts.

1.2.5   Electrostatic Generators

In electromagnetic generators, current carrying conductors are moved against the electromagnetic forces acting upon them. In electrostatic generators, movement of electrically charged particles occurs against the electrostatic forces acting upon them.

The working principle of an electrostatic generator will be explained using Fig. 1.27 as illustration. An insulated belt of width b with charge carrier density σ is suspended in the electric field $E(x)$ between two electrodes with spacing s. The charge on a strip of height dx is given by:

$$dq = \sigma .b\ dx.$$

The force acting upon the entire belt is :

$$F = \int_{o}^{s} dF = \int_{o}^{s} E(x)dq = \sigma.b \int_{o}^{s} E(x)dx.$$

If the belt is moved with constant velocity v = dx/dt against this force, the necessary mechanical power is:

$$P = F.v = \sigma.b.v \int_{o}^{s} E(x)dx.$$

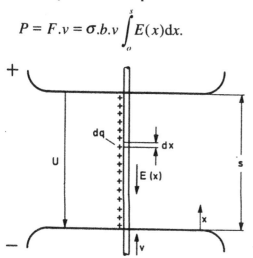

Fig 1.27 For illustration of the working principle of electrostatic generators

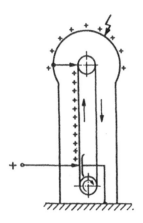

Fig 1.28 Belt generator after van de Graaff

Since

$$I = \frac{dq}{dt} = \sigma.b.v \text{ and } U = \int_o^s E(x)dx$$

it can be seen that the mechanical power required for the drive is equal to the electrical power output IU.

The most common type of electrostatic generator is the belt generator devised in 1931 by *R.J. van de Graaff*, the working principle of which is illustrated in Fig. 1.28. An insulated belt is run over rollers and electrostatically charged by an excitation arrangement. A strongly inhomogeneous electrode configuration is used for this purpose; charge carriers formed by collision ionization at the sharp electrode are trapped by the belt on their way to the opposite electrode. A similar arrangement of the high-voltage end serves to discharge the belt. If another excitation arrangement of the opposite polarity is provided at the high-voltage level for the downwards moving side of the belt, twice the amount of current is obtained.

Belt generators of the pressure tank type have already been constructed for voltages up to 25 MV, where currents less than 1 mA are usual [*Herb* 1959]. Instead of the insulated belt, one can also use highly insulating liquids or dust-like solid materials as carriers of the electric charge.

For voltages up to some 100 kV, diverse types of electrostatic machines with drum or disc-shaped rotors have been built. Among some of the advantages of these machines are good control over the constancy of the output voltage, as well as low self-capacitance, which results in an essentially safe high-voltage device [*Felici* 1957]. Belt generators are predominantly used in physical research applications; they are seldom seen in testing applications due to their low current loading capability.

Measurement of High Direct Voltages[6]

1.2.6   Measurement with High-Voltage Resistors

The measurement of a direct voltage can, with the aid of resistors, be reduced to the measurement of a direct current. Basically, two circuits shown in Fig 1.29 are possible.

In high-voltage applications, there is the problem that the measuring current must be chosen to be very small, of the order of 1 mA for example, because of the permitted loading of the voltage source and heating of the measuring resistor. A small current is however easily falsified by error currents; these occur in the form of leakage currents in insulating materials and on insulating surfaces, and also as a result of corona discharges. Certain details of the design of high-voltage measuring resistors shall be given in section 2.4.1.

The characteristic parameter of the direct voltage measured depends on the working principle of the ammeter at earth potential, and connected in

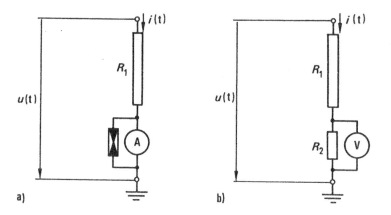

Fig. 1.29   Measurement of direct voltage
a) with current limiting resistor, b) with a resistive divider

---

[6] Comprehensive treatment, among others, in *Böning* 1953; *Craggs, Meek* 1954; *Sirotinski* 1956; *Paasche* 1957; *Schwab* 1969; *Kuffel, Zaengl* 1984.

series with the measuring resistor. A sensitive moving-coil instrument is usually chosen, the indication of which is a measure of the arithmetic mean value $\bar{U}$ of the direct voltage (Fig.1.29a). The measuring range is easily changed in any case by parallel connection of a resistance R_2 to the measuring instrument, which turns the series resistor into a resistive voltage divider. Even electronic sensors with very low measuring uncertainty (error) are also in use [*Marx , Zirpel* 1990]. Instead of the ammeter, a voltmeter with an internal resistance preferably much larger than R_2 may also be connected (Fig.1.29b). For high voltages (above 500 kV) the high-voltage resistor should be capacitively graded in order to achieve an uniform voltage stressing in case of breakdowns on the specimen. This can be achieved by appropriate construction of the resistor or with concentrated capacitances (a few 10 pF). If the grading is effected with capacitors, the d.c divider can be extended to function as a parallel mixed RC-divider which also enables the measurement of the ripple factor.

1.2.7   Measurement of R.M.S. Value by Means of Electrostatic Voltmeters

As may be seen from the description, in section 1.1.14, of the working principle of electrostatic voltmeters, this type of instrument can also be used for direct voltages. Electrostatic voltmeters do in fact represent the best way of measuring high direct voltages directly. It is a question of a loss-free measurement which can also be performed when no current may be drawn from the voltage source.

In this method, voltage measurement is reduced to measurement of a field strength at an electrode, which is particularly illustrated by the arrangement indicated in Fig. 1.16b. For high direct voltages space charges will occur when electrodes of small radius of curvature are used and the system is not fully screened. These space charges, or surface charges adhering to the surface of insulating materials, can affect the field strength at the rotating electrode segment and so result in considerable error.

1.2.8   Voltmeter and Field Strength Meter Based upon the Generator Principle

Consider the electrode arrangement shown in Fig.1.30a, where a measuring electrode of area A, assumed to be at earth potential, has constant surface charge density $\varepsilon_0 E$ produced by the steady field strength E. The total charge on the measuring electrode is given by:

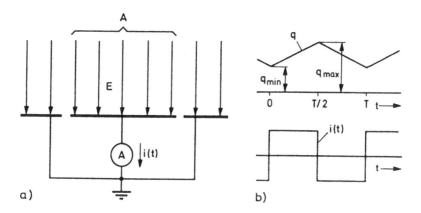

Fig. 1.30 Measurement of voltage and field strength according to generator principle
a) schematic measuring arrangement, b) charge and current curves

$$q = \int_{(A)} \varepsilon_0 \, E dA = \varepsilon_0 \, AE.$$

The charge q is now allowed to vary between the values q_{max} and q_{min} as shown in Fig.1.30b, this being done by periodic covering and uncovering of a portion of the measuring electrode by an earthed plate. An alternating current $i(t) = dq/dt$ then flows in the earth lead; the curves of the positive and negative half - periods are the same if the covering and uncovering movement is uniform. The arithmetic mean value of the current between two zero-crossovers is then:

$$\frac{1}{T/2} \int_0^{T/2} \frac{dq}{dt} \, dt = \frac{2}{T}(q_{max} - q_{min}).$$

For rectification, this value corresponds to the arithmetic mean value taken $\bar{I}$ over a whole period. If the measuring electrode is completely covered at $t = 0$, q_{min} would be zero, and we have :

$$\bar{I} = \frac{2}{T} q_{max} = \frac{2}{T} \, \varepsilon_0 \, AE.$$

Thus, $\bar{I}$ is proportional to the field strength and can be used to measure the latter. If the frequency of the mechanical movement is high, even low steady field strengths can be measured well because of the correspondingly

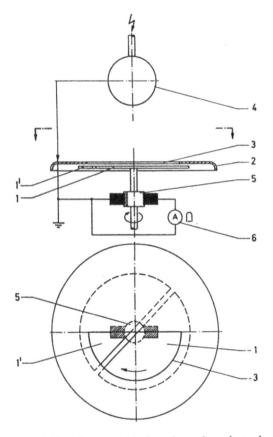

Fig. 1.31 Voltmeter with the sphere-plate electrode configuration
1,1´ Revolving semicircular discs, 2 Semicircular opening, 3 Earthed covering plate,
4 High-voltage electrode, 5 Commutator, 6 Ammeter

high dq/dt. This principle was indeed applied for the first time in 1926, by
A. Matthias and *H. Schwenkhagen* in thunderstorm investigations, to measure
electric field strengths at ground level. A different type of field strength
meter, instead of covering the electrode, uses an oscillatory movement of
the measuring electrode in the field direction to generate the alternating
current $i(t)$.

Using the arrangement shown schematically in Fig. 1.31 as an example,
we shall show how a direct voltage U may be measured according to the
same principle [*Kind* 1956]. The two measuring electrodes 1 and 1´ are
alternately passed underneath the semicircular opening 2 of the earthed
plate 3 by the drive; this produces a partial capacitance (varying between
zero and a maximum value), between each of the measuring electrodes and

the high voltage electrode 4. At constant rate of revolution, therefore, a periodic alternating current $i(t)$ flows between the measuring electrodes, which is rectified by the commutator 5. The arithmetic value $\bar{I}$ after rectification can be recorded by the moving-coil ammeter 6. Owing to the proportionality of the field strength E at the electrodes and the voltage U to be measured, $\bar{I}$ is proportional to U. If, in order to determine the surface charge, we introduce the maximum value C_m of the periodically varying partial capacitance between one measuring electrode and the high-voltage electrode, we have :

$$q_{max} = C_m U \quad \text{and} \quad q_{min} = 0$$

and it follows that :

$$\bar{I} = \frac{2}{T} C_m U.$$

The principle just described has been applied in various different ways [*Prinz* 1939; *Schwab* 1969].

1.2.9   Measurement of High Direct Voltages with Rod - Gaps

The breakdown voltage of sphere-gaps under direct voltages has a larger spread (~5%) than under alternating voltages; furthermore, abnormally low breakdown voltages, down to 50% value, can be caused by the unavoidable presence of dust particles. In contrast, rod-gaps indicate a very small spread of the breakdown voltage. They are therefore included in the specifications (IEC Publ. 60-1, 1989) for measurement of high direct voltages.

Rod-gaps can be set-up either vertically or horizontally (Fig.1.32). The breakdown voltage of a rod-gap referred to standard conditions increases linearly with the gap spacing and, with appropriate construction, is independent of polarity and the surroundings. The linear increase of breakdown voltage with the gap spacing can be explained by the stable streamer pre-discharge occurring at both the electrodes, which from a specific voltage onwards (ca. 150kV), becomes stable before the breakdown. The breakdown voltage $\bar{U}_{do}$ can be calculated from the expression:

$$\bar{U}_{do} = 2 + 5.34 \cdot s$$

($\bar{U}_{do}$ in kV, s in cm). The breakdown voltage $\bar{U}_{do}$ as a function of spacing s is represented in Fig.1.32c.

The actual breakdown voltage $\bar{U}_d$, with relative air density d as per 1.1.10 and the humidity factor k works out to:

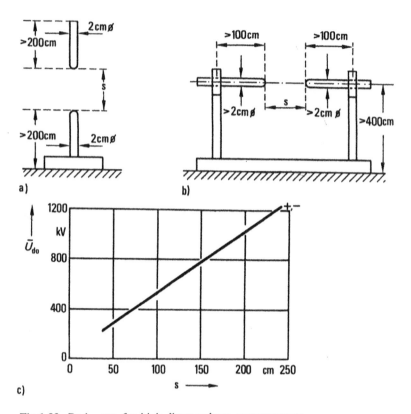

Fig.1.32 Rod - gap for high direct voltage measurements
a) horizontal arrangement, b) vertical arrangement, c) breakdown voltage U_{do} of rod-gap for variable gap spacing s

$$U_d = d.\ k.U_{do}$$

With absolute humidity h_a and $h_{a0} = 11$ g/m^3 [*Peschke* 1968], we have:

$$k \approx 1 + 0.014\ (\ h_{a0} - h_a\ ).$$

Compared to the sphere-gap, while referring to the standard conditions, additionally the relative humidity is also to be accounted for in rod-gaps. The measuring uncertainty is lesser than ±3%. The range of application of the rod-gap has been experimentally established for spacings between 30 cm and 250 cm or 150 kV and 1300 kV. The absolute humidity shall be lower than 13 g/m^3. The tips of round rods shall be hemispherical and those of square rods shall be blunt. Brass, copper or aluminium can be used as material for the rods. Loading of the voltage source up to breakdown is negligible (< 1mA) [*Feser, Hughes* 1988].

1.2.10 Other Methods for the Measurement of High Direct Voltages

The method of measuring alternating voltages using sphere-gaps, described in 1.1.10, is also suitable for the determination of the peak value $\hat{U}$ of high direct voltages.

Fundamentally different methods for the measurement of high direct voltages have been developed for special cases of application in physics. Those methods which allow the measured quantity to be expressed in terms of base units and of accurately known fundamental constants, are of particular scientific significance. For example, to calibrate the voltage measuring devices of elementary particle accelerators, protons are accelerated in an electric field which is proportional to the voltage to be measured. At certain kinetic energies of these protons, on collision with light atomic nuclei, resonant nuclear transformations occur which permit very exact determination of the applied direct voltage [*Jiggins* , *Bevan* 1966; *Oechsler*, 1991].

1.2.11 Measurement of Ripple Factor

High-ohmic resistance dividers do not normally have the bandwidth to measure the ripple factor correctly. They measure the mean value of the direct voltage. In contrast, the mixed voltage divider described in 1.2.6 is suited for the measurement of the direct voltage $\bar{U}$ and the ripple voltage δU.

In case measuring sensitivity requires, very small ripple voltages can be also measured with a separate measuring circuit which enables a direct measurement of the temporal nature of the voltage $u(t) — \bar{U}$.

Fig. 1.33 shows a simple circuit in which a high-voltage capacitor C separates the ripple from the direct voltage. The voltage divider made up

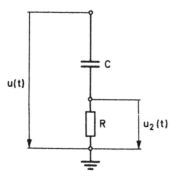

Fig.1.33 Circuit for the measurement of ripple factor

of C and R has a divider ratio of zero for d.c. voltages; on the other hand, for alternating voltage with angular frequency ω, we have

$$\frac{\underline{U}_2}{\underline{U}} = \frac{jR\omega C}{1 + jR\omega C}.$$

Now if the condition

$$u_2(t) \approx u(t) - \overline{U}$$

is to be well satisfied, the divider ratio must be as near to 1 as possible for all frequencies in the ripple spectrum, which is the case when

$$R\omega C \gg 1.$$

This is easily fulfilled for the fundamental frequency ω and so faithful reproduction of the ripples is assured. During such measurement of the ripple factor, it must be borne in mind that in case of an eventual breakdown of the specimen, a very high voltage appears across the resistance R.

1.3 Generation and Measurement of Impulse Voltages

Impulse voltages are required in high-voltage tests to simulate the stresses due to external and internal overvoltages, and also for fundamental investigations of the breakdown mechanisms. They are usually generated by discharging high-voltage capacitors through switching gaps onto a network of resistors and capacitors, whereby voltage multiplier circuits are often used. The peak value of impulse voltages can be determined with the aid of measuring gaps, or better, be measured by electronic circuits combined with voltage dividers. The most important measuring devices for impulse voltage are, however, the cathode-ray oscilloscope and the digital recorder, which allow the complete time characteristic of the voltage to be determined by means of voltage dividers.

1.3.1 Characteristic Parameters of Impulse Voltages

In high-voltage technology a single, unipolar voltage pulse is termed an impulse voltage; three important examples are shown in Fig. 1.34, with reference to the possible characteristic parameters. The time dependence, as well as the duration of the impulse voltage, depend upon the method of generation. For basic experiments, rectangular impulse voltages are often

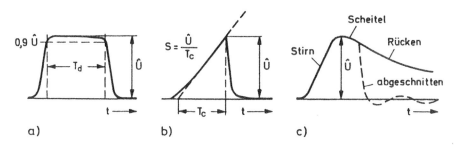

Fig. 1. 34 Examples for impulse voltages
a) rectangular impulse voltage
b) wedge-shaped impulse voltage
c) double exponential impulse voltage front peak tail chopped

used which rise abruptly to an almost constant value, as well as wedge-shaped impulse voltages characterized by a rise which is as linear as possible up to breakdown, and described simply by the steepness S. For testing purposes, double exponential impulse voltages have been standardized; without appreciable oscillation these rapidly reach a maximum, the peak value $\hat{U}$, and finally drop less abruptly to zero. If an intentional or unintentional breakdown occurs in the high-voltage circuit during the impulse, leading to a sudden collapse of the voltage, this is then called a chopped impulse voltage. The chopping can occur on the front, at the peak or in the tail section of the impulse voltage. The transient phenomenon thereby induced is mainly responsible for the oscillations indicated in Fig. 1.34c.

For overvoltages following lightning strokes, the time required to reach the peak value is of the order of 1 μs; they are named atmospheric or external overvoltages. Voltages generated in a laboratory to simulate these are called lightning impulse voltages. For internal overvoltages, occurring as a consequence of switching operations in high-voltage networks, the time taken to reach the peak value is at least about 100 μs. Their reproduction in the laboratory is effected by switching impulse voltages; these are of approximately the same shape as lightning impulse voltages, but last considerably longer.

In the case of impulse voltages for testing purposes, the shape of the voltage is determined by certain time parameters for the front and tail, as shown in Fig. 1.35 (VDE 0432-1; IEC Publ.60-1).

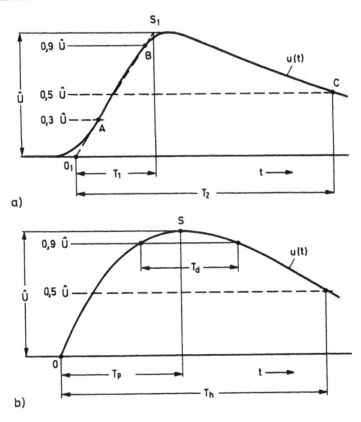

a)

b)

Fig. 1.35 Characteristic parameters of standard test impulse voltages
a) lightning impulse voltage, b) switching impulse voltage

Since the true shape of the front of lightning impulse voltages is often difficult to measure, the straight line $O_1 - S_1$ through the points A and B is introduced as an auxiliary construction on the front, to characterize the latter. Then the time T_1 to front, as well as the time T_2 to half-value, being the time from O_1 to point C, are also determined. In general, lightning impulse voltage of shape 1.2/50 are used, which means an impulse voltage with $T_1 = 1.2$ μs ±30 % and $T_2 = 50$ μs ±20 %. On the other hand, recording the much slower switching impulse voltage presents no difficulties; hence the true origin 0 and the true peak S can be utilized for standardization. For tests with switching Impulse voltages, the shape 250/2500 is often used, which corresponds to $T_p = 250$ μs ±20 % and $T_h = 2500$ μs ±60 % (T_p = time to peak, T_h = time to half-value). While testing transformers, in order to denote the duration of the switching impulse voltage, the time T_d

during which the instantaneous value of the voltage lies above $0.9\hat{U}$ is also often quoted instead of T_h.

The curves of lightning impulse voltages often have high-frequency oscillations superimposed, the amplitude of which may not exceed $0.05\ \hat{U}$ in the region of the peak. It is assumed in this case that the frequency of the oscillations is at least 0.5 MHz, otherwise the actually observed maximum value of the voltage is taken as the peak value of the lightning impulse voltage .Oscillations on the front up to half the amplitude of the lightning overvoltage shall not exceed an amplitude of $0.5\hat{U}$.

Generation of Impulse Voltages

1.3.2 Capacitive Circuits for Impulse Voltage Generation[7]

Fig. 1.36 shows the two most important basic circuits, denoted "circuit a" and "circuit b", used for the generation of impulse voltages. The impulse capacitor C_s is charged via a high charging resistance to the direct voltage U_0 and then discharged by ignition of the switching gap F. The desired impulse voltage $u(t)$ appears across the load capacitor C_b. The circuits a and b differ from one another in that, in the one case, the discharge resistor R_e is connected in front of, and in the other, behind the damping resistor R_d.

The value of the circuit elements determines the curve shape of the impulse voltage. The basic working principle of both circuits can be readily understood from the following simple considerations. The short time to front requires rapid charging of C_b to the peak value $\hat{U}$, and the long time to tail, a slow discharge. This is achieved by $R_e \gg R_d$. Immediately after

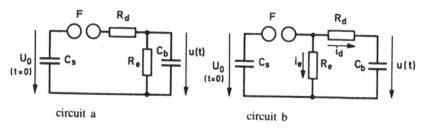

circuit a   circuit b

Fig. 1.36 Basic diagrams of impulse voltage circuits

---

[7] Comprehensive treatment among others in *Craggs, Meek* 1954; *Strigel* 1955; *Widmann* 1962; *Helmchen* 1963; *Beyer et al.* 1986.

ignition of F at $t = 0$, almost the full charging voltage U_0 appears across the series combination of R_d and C_b in both circuits. The smaller the value of the expression $R_d C_b$, the faster is the rate at which the voltage $u(t)$ reaches its peak value. The peak value $\hat{U}$ cannot be greater than is determined by distribution of initially available charge $U_0 C_s$ onto $C_s + C_b$. For the utilization factor η therefore we have :

$$\eta = \frac{\hat{U}}{U_0} \leq \frac{C_s}{C_s + C_b}.$$

Since for a given charging voltage $\hat{U}$ should generally be as high as possible, one will choose $C_s \gg C_b$. The exponential decay of the impulse voltage on the tail would then, in circuit a, occur with the time constant $C_s(R_d + R_e)$, and in circuit b with the time constant $C_s R_e$. The impulse energy transformed during a discharge is then :

$$W = \frac{1}{2} C_s U_0^2 .$$

If the highest possible charging voltage is substituted for U_0 in this expression, we obtain the maximum impulse energy as an important characteristic parameter of the impulse voltage generator.

In the above explanation of the operating mode of the circuits, it was assumed that at $t = 0$ the impulse capacitors C_s were charged to a voltage U_0. U_0 is the value of the charging voltage at which F breaks down, either by itself or by means of an auxiliary discharge. Thus, for self-triggered operation, an increase in the peak value of the impulse voltage $\hat{U}$ can only be achieved by increasing the spacing of F. Merely increasing the direct voltage applied in front of the charging resistor would only result in C_s charging up faster to the value U_0, and F breaking down spontaneously in shorter intervals of time. Hence the impulse rate would increase and not the amplitude of the impulse voltage generated.

For given d.c. charging voltage, to obtain impulse voltage with as high a peak value as possible, the multiplier circuit proposed by E. Marx in 1923 is commonly used. Several identical impulse capacitors are charged in parallel and then discharged in series, obtaining in this way a multiplied total charging voltage, corresponding to the number of stages. The mechanism of the Marx circuit will be explained with the aid of the impulse generator shown in Fig. 1.37 with n = 3 stages in circuit b connection. The impulse capacitors of the stages C_s' are charged to the stage charging voltage U_0', via the high charging resistors R_L' in parallel.

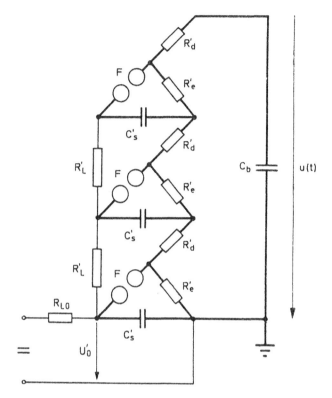

Fig. 1.37 Multiplier circuit after Marx for 3 stages in circuit b connection

When all the switching gaps F break down, C_s' will be connected in series, so that C_b is charged via the series connection of all the damping resistors R_d'; finally, all C_s' and C_b will discharge again via the resistors R_e' and R_d'. It is expedient to choose $R_L' \gg R_e'$. The n-stage circuit can be reduced to a single stage equivalent circuit, such as circuit b, where the following relationships are valid:

$$U_0 = n . U_0' \qquad R_d = n . R_d' \qquad C_s = \frac{1}{n} C_s' \qquad R_e = nR_e' .$$

For the operation of the Marx circuit it is essential that all the switching gaps F, which are normally sphere-gaps of adjustable spacing, break down almost simultaneously. This is usually achieved by setting the lowest sphere gap to a slightly smaller spacing or triggering it first by an auxiliary discharge. Transient overvoltages occur due to the breakdown of the lowest sphere-gap on the upper sphere-gaps which therefore break down one after the other. The overvoltages depend on the construction of the impulse generator (stray capacitances to earth and between stages, inductances of the stages), the switching elements (damping resistance R_d', discharge resistance R_e') and the number of stages [*Rodewald* 1969; *Heilbronner* 1971]. In any construction of a multi-stage impulse generator, the trigger range decreases with increase of the internal discharge resistance R_e'. An increase of the damping resistance R_d' is necessary for generating switching overvoltages; a decrease of the discharge resistance R_e' results in an increase of the energy (larger capacitance C_s') of an impulse generator for lightning impulses. By appropriate switching elements and measures [*Rodewald* 1974; *Feser* 1974] or by triggered operation of all the sphere-gaps [*Bishop, Feinberg* 1971; *Feser* 1974] it can be ensured that a sure triggering of multi-stage impulse generators is guaranteed even in such cases. With the robust, simple sphere-gap, a triggering range of more than 20% is achievable in the most unfavourable conditions even if the sphere-gap in the first stage only is triggered by a trigger pulse. The natural overvoltages lead to a firing of the entire impulse generator - a 10 stage impulse generator completely firing in a few 100 ns[*Heilbronner* 1971].

Impulse voltage generators have already been built for voltages up to 10 MV and for impulse energies of a few 100 kWs, where the charging voltages per stage are usually of the order of 100 ... 300 kV. The utilization factor η depends on the shape of the impulse voltage to be generated and generally lies between 0.7 and 0.9. It is also principally higher for circuit b than for circuit a, especially for impulse voltages with comparatively shorter times to tail or in the case of small load capacitances.

Important construction elements for a multi-stage impulse generator are the capacitors, whose construction decides the type of generator. In impulse generators for testing purposes, the stages are arranged either vertically above one another or in a zigzag manner. The sphere-gaps mostly lie above one another with mutual visual connection. Important for practical testing is the arrangement of the resistors, which should be easily interchangeable with the help of ladders or working platforms. Due to safety reasons, all capacitors will be short-circuited while interchanging the resistors. Multi-stage generators can be operated with reduced number of stages (lower inductance, higher impulse capacity) or in a parallel/series connection of stages.

1.3.3   Calculation of Single-Stage Impulse Voltage Circuits

For the design of impulse voltage circuits it is necessary to establish relationships between the values of the circuit elements and the characteristics of the voltage shape. Because of the higher utilization factor, impulse generators are built predominantly in the basic circuit b connection. For this reason the impulse shape has been calculated in Appendix 4.3 for this circuit, using the symbols shown in Fig. 1.36b. For the impulse voltage curve the solution is :

$$u(t) = \frac{U_0}{R_d C_b} \cdot \frac{\tau_1 \tau_2}{\tau_1 - \tau_2} \left(e^{-t/\tau_1} - e^{-t/\tau_2}\right).$$

It is seen that the impulse voltage is given by the difference of two exponentially decaying functions with time constants τ_1 and τ_2. Fig 1.38 shows the curve which reaches the peak value $\hat{U}$ at time T_p.

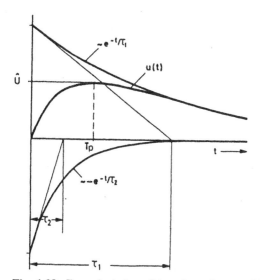

Fig. 1.38 For calculation of impulse voltages with double exponential shape.

With the usually satisfied approximation

$$R_e C_s >> R_d C_b,$$

the following simple expressions are obtained for circuit b:

$$\tau_1 \approx R_e(C_s + C_b) \; ; \; \tau_2 \approx R_d \frac{C_s C_b}{C_s + C_b}; \; \eta \approx \frac{C_s}{C_s + C_b}.$$

For circuit a of Fig. 1.36a, the same general solution is true, but with:

$$\tau_1 \approx \left(R_d + R_e\right)\left(C_s + C_b\right) \; ; \qquad \tau_2 \approx \frac{R_d R_e}{R_d + R_e} \cdot \frac{C_s C_b}{C_s + C_b} \; ;$$

$$\eta \approx \frac{R_e}{R_d + R_e} \cdot \frac{C_s}{C_s + C_b} \; .$$

The impulse shape is described uniquely by τ_1 and τ_2. Consequently the characteristics as in Fig. 1.35 must also be functions of τ_1 and τ_2. Since in general $\tau_1 \gg \tau_2$, the condition quoted for simplified calculation of τ_1 and τ_2 from the circuit elements is then also satisfied. For lightning impulse voltages of the standard form 1.2/50, we have:

$\tau_1 = 68.22\ \mu s$, $\tau_2 = 0.4050\ \mu s$.

A lightning impulse voltage with the shortest possible time to front

$T_1 = 0.7 \cdot 1.2 = 0.84\ \mu s$

and the longest possible time to tail

$T_2 = 1.2 \cdot 50 = 60\ \mu s$

often poses a critical case for the transfer performance. For such a voltage waveform 0.84/60, we have:

$\tau_1 = 83.67\ \mu s$, $\tau_2 = 0.2746\ \mu s$.

For the characteristics of switching impulse voltages, we have :

$$T_p = \frac{\tau_1 \tau_2}{\tau_1 - \tau_2} \ln \tau_1 / \tau_2$$

$$T_h \approx \tau_1 \ln \frac{2}{\eta} \quad \text{for } T_h \geq 10 T_p .$$

If the conditions given above are only partly fulfilled, then the general solution for $u(t)$ must be evaluated.

The shape of the voltage for lightning impulse voltages often deviates considerably from the theoretically calculated, particularly on the front and at the peak. Reasons for these are the firing of the impulse generator, for the oscillations on the front of the lightning impulse voltage, and the unavoidable inductances of the elements and the spatial construction of the generator for the oscillations at the peak of the lightning impulse voltage. Fig.1.39 shows an example of the voltage wave form with superposed oscillations.

The front-oscillations, which render an unambiguous determination of the 30% point difficult, result from the rapid firing of the upper stages of a multi-stage generator. A voltage is suddenly coupled through longitudinal capacitance of the generator stages to the connecting lead of the load capacitance , which gets reflected at that end. By means of the damping resistance between the impulse generator and the load capacitance, this

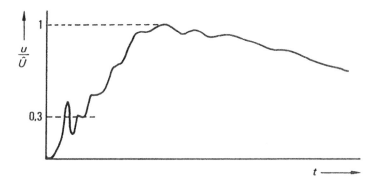

Fig. 1.39 Front-oscillation and oscillations at the peak of lightning impulse voltage

oscillation can be appreciably reduced. The superposed oscillations at the peak of the lightning impulse voltage are caused by the circuit inductance L which can be assumed to be in series with R_d in the equivalent circuit. In order to prevent the disturbing oscillations, the circuit must be aperiodically damped;

$$R_d \geq 2\sqrt{L\frac{C_s + C_b}{C_s C_b}}.$$

In setups for high voltages (large inductance L) or while testing large load capacitances C_b, it is often difficult to fulfill this condition since the front time is also to be restricted to within permissible tolerances $(T_1 \sim R_d \cdot C_b)$.

1.3.4 Generation of Lightning Impulse Voltages or Switching Impulse Voltages under Capacitive Loads

Most of the specimen in high-voltage technology represent a capacitive load to the voltage source and can be included in the value of the load capacitor C_b. In industrial testing practice, one would have a certain impulse generator e.g. 2 MV, 400 kWs and the question that arises is, with which and with how many elements a standard impulse waveform can be generated?

Fig. 1.40 shows the dependence of the usually still-freely choosable elements of the equivalent circuit R_e and R_d (circuit b, Fig. 1.36) on the load capacitance C_b. It may be recognised that for the generation of the lightning impulse voltage within the standardised limits for an usual loading range upto 5 nF, only one discharge resistance R_e, but more number of

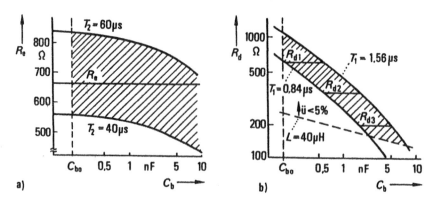

Fig. 1.40 Tolerance regions for standard lightning impulse voltages, as a function of the capacitive load C_b (Example: Impulse generator 2 MV, 400 kWs)
a) tail time T_2,   b) front time T_1

damping resistances R_d are required. If the damping resistance R_d is too small, or the load capacitance C_b too large, the standard impulse voltage wave form is constrained by the overshoot. The inductance $L_g{'}$ of multi-stage generators can be estimated to be $3\mu H/stage$. The total inductance L is obtained from the generator inductance ($L_g = n.L_g{'}$) and the inductance of the external circuit (including the specimen), which can be estimated to be approximately $1\mu H/m$.

For generating switching impulse voltages, comparable conditions exist. It means that as a rule, one needs 2 discharge resistances (for 50 μs and 2500 μs) as well as a large number of damping resistances R_d; the arrangement of the resistances is possible either in the generator itself or outside [*Feser* 1974]. Arranging the resistances externally simplifies the firing of the impulse generator, but requires too much space at high voltages. Arrangement of resistances within the generator is therefore preferred.

1.3.5   Generation of Lightning Impulse Voltages under Low Inductive Loads (Transformers)

It is often difficult to realize the minimum tail time of 40 μs while testing low inductances with lightning impulse voltages. Fig. 1.41a shows the simple equivalent circuit, Fig. 1.41b the basic waveform of the lightning impulse voltage without the inductance L_b (curve 1) and with a low inductance L_b (curve 2). Shown in Fig. 1.41c is the minimum inductance L_b that just enables generation of a standard lightning impulse voltage as a function of the impulse capacitance C_s.

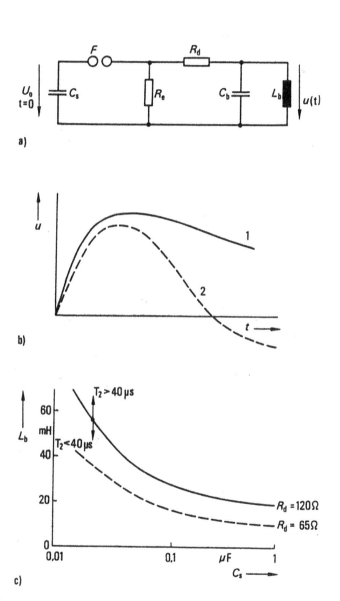

Fig. 1.41 Generation of standard lightning impulse voltages under inductive loads
a) equivalent circuit (circuit b, R_e - discharge resistance for switching impulse voltages)
b) voltage waveforms, 1 without L_b, 2 with L_b
c) required impulse capacitance C_s ($C_b \sim 10$ nF)

It may be recognised that the real limitation is from the damping resistance R_d which has to have a specific value in order to damp the oscillations as well as to obtain a time to front according to the standards. By increasing the impulse capacitance C_s through parallel connection of stages, one can conduct practical tests on inductances $L_b > 4$ mH with standard lightning impulse voltage. But still lower inductances are encountered while testing the low voltage winding of 3 phase transformers. For example, in a 525/24 kV, 3 phase transformer with a reactive power rating of 400 MVA and a short-circuit voltage of 15%, the inductance L_{bHS} of the high-voltage winding is 330 mH and the inductance L_{bNS} of the low-voltage winding is 0.7 mH. Testing of the high-voltage winding with lightning impulse voltages poses therefore no problems from the point of view of the voltage waveform, whereas while testing the low-voltage winding the time to half-value of 40 μs cannot be obtained either with circuit a or circuit b. The cause therefor lies in the current limitation by R_d [Kannan, Narayana Rao 1973; Feser 1978]. If R_d is bridged by an inductance L_d , the time to half-value can be extended. In such a case voltage division occurs on the tail portion between L_d and L_b during the discharge of C_s. For very low inductances L_b, this voltage division must be ensured even during the front portion of the lightning impulse voltage by the additional parallel resistance R_p.

The corresponding equivalent circuits are shown in Fig. 1.42 . The additional impulse circuit elements L_d and R_p can be dimensioned according to the following criteria: the inductance L_d should be so designed that its impedance has a negligible effect during the rising portion of the lightning impulse voltage and practically short-circuits the damping resistance during the tail portion of the lightning voltage. Due to voltage division, L_d must be smaller than L_b. An inductance L_d of 400 μH is adequate for testing low inductances (< 40 mH). For very low inductances $L_b < 1$ mH, the inductance L_d shall be < 100 μH ; additionally, one requires the parallel resistance $R_b = L_b.R_d/L_d$. Since low inductances L_b are encountered in practice especially when testing the low voltage windings of transformers, only one stage of a multi-stage impulse generator needs to be provided with these additional elements.

1.3.6 Chopped Lightning Impulse Voltages

Chopped lightning impulse voltage is intended for the testing of the turn - to- turn insulation of windings e.g., in transformers. Due to the rapid voltage collapse, a non-linear voltage distribution results along the winding [Kind,

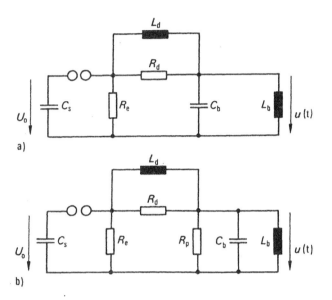

Fig. 1.42 Impulse circuits for generation of lightning impulse voltages under inductive loads
a) for inductances 4 mH < L_b < 40 mH
b) for extremely low inductances (L_b < 4 mH)

Kärner 1982]. In testing practice, such rapid voltage collapse can be obtained for voltages up to about 600kV from a sphere-gap and for higher voltages by means of multi-stage chopping gaps connected in parallel to the specimen. The often applied rod-gap is not so well suited as a chopping gap at high voltages due to its slow voltage collapse (~ 100 ns).

Fig. 1.43 shows a circuit diagram of a multiple chopping gap with capacitive grading of the series connected individual gaps [*Feser, Rodewald* 1972]. The bottommost stage can be fired by controlled triggering such that the chopping point can be freely chosen and is thus reproducible e.g., at the peak or on the tail portion of the lightning impulse voltage. The diagonal resistances lead to increasing overvoltages in the successive stages and thereby to a rapid firing of the generator. If a portion of the overvoltages arising from the firing of the stages is tapped across the resistance R_T, this overvoltage can be utilised to trigger the immediately following stage sphere -gap, thereby reducing the spread in the chopping time considerably. The series connection of capacitances C results in an uniform voltage distribution across the stage sphere-gaps just prior to the chopping point. By appropriate dimensioning, a grading column can also serve as the load capacitance.

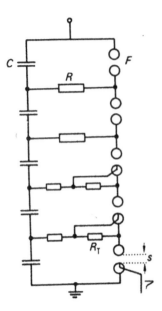

Fig. 1.43 Circuit diagram of a 5 stage multiple chopping gap

The multiple chopping gap is adjusted to the required amplitude of the voltage to be chopped, by short-circuiting of some stages (as a rule, 200kV steps) or by varying the spacings of the stage gaps.

1.3.7   Generation of Oscillatory Impulse Voltages

In practice, lightning impulse voltages as also switching impulse voltages often occur as oscillatory voltages. Most important advantage while generating oscillatory impulse voltages is the higher utilisation factor that results especially under capacitive load. Fig. 1.44 shows the simple equivalent circuit for generation of oscillatory lightning impulse voltages or switching impulse voltages. It may be recognised that the damping resistance R_d is replaced by an inductance L_d.

For the no-damping case ($R_e = \infty$), we have :

$$u(t) = U_0 \cdot \frac{C_s}{C_s + C_b}(1 - \cos\alpha t) \quad \text{with } \alpha = \sqrt{\frac{1}{L_d}\frac{C_s + C_b}{C_s \cdot C_b}}$$

In the ideal case i.e., with very low losses in the inductance L_d and the capacitors and for $C_s \gg C_b$, the utilisation factor $\eta \approx 2$. In practice, one

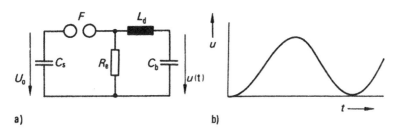

Fig. 1.44 Generation of oscillatory impulse voltages in a series resonance circuit
a) equivalent circuit, b) voltage waveform

achieves with a charging voltage of U_0 an output voltage of nearly $2.U_0$.
The time to peak value of the impulse voltage works out approximately to:

$$T_p \approx \pi \sqrt{L_d . C_b} .$$

Oscillatory impulse voltages are preferably chosen for the *in-situ* testing
of e.g., SF_6 setups[*Kind* 1974; *Feser* 1981]. For testing generators and
motors, symmetrical oscillatory impulse voltages are used [*Schuler* 1980].
Fig. 1.45 shows the simple equivalent circuit. It is seen that the voltage
at the specimen (L_b) raises rapidly and dies down oscillating at the frequency
resulting from C_x and L_b. The rise of the oscillatory impulse voltage is
governed by the time constant given by the loss resistances (of capacitor,
sphere-gap and connecting leads) that are always present and the inductor
input capacitance C_b; it is of the order of a few 10 ns. The winding (turn-
to-turn) insulation of generators can be stressed closest to practice using
this test circuit and tested.

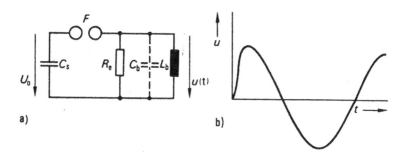

Fig. 1.45 Generation of oscillatory impulse voltages in a parallel resonance circuit
a) equivalent circuit, b) voltage waveform

1.3.8   Generation of Steep Impulse Voltages

Generation of pulses with rise times in the ns range requires special circuits and constructional elements. Double exponential voltage pulses e.g., a pulse with $T_1 = 5$ ns and $T_2 = 200$ ns is generated with special switching elements which are additionally connected either to a direct voltage generator (for voltages below 300 kV) or to an impulse generator (for voltages above 300 kV). Fig.1.46a shows the simple equivalent circuit and Fig.1.46b the functioning of a circuit under pulse energization. For generating steep pulses the additional circuit consisting of the capacitor C_N and the sphere-gap F_N must be low inductive i.e., must be very compact in construction. The rise time T_a of the steep pulse is primarily determined by the stray parameters of the circuit set-up and the breakdown time of the sphere gap F_N [*Kärner* 1967].

In practice, the sphere-gap is realized as an encapsulated sphere-gap with as homogenous a field as possible and SF_6 as insulating gas. In many applications, the additional circuit is supplemented by a strip - line antenna which can be represented by an ohmic resistance [*Feser* 1987]. A pulsed electromagnetic field is then generated in the strip-line, which can be used e.g., for EMI tests on electrical measuring equipment.

Short rectangular impulse voltages can be generated quite well with energy storage devices of the transmission line type. In a much used setup, a high-

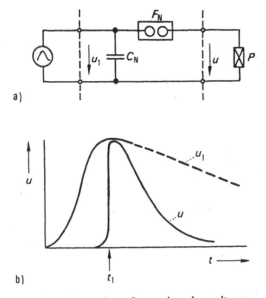

Fig.1.46   Generation of steep impulse voltages
a) equivalent circuit,   b) voltage waveform

voltage cable is charged to a direct voltage U_0 via a high resistance and then discharged through a sphere-gap on to an initially uncharged cable, at the end of which the test object is connected. The duration of the voltage impulse which develops across the test object is twice the travelling wave transit time of the charging cable; the peak value depends upon the impedance of the test object and is, at best, equal to U_0. In a different circuit, high-voltage capacitors are switched on to a delay cable short-circuited at the end, the effective length of which can easily be varied to obtain impulse voltages of diverse durations [*Winkelnkemper* 1965].

Steep impulse voltages with rise-times of a few ns appear in circuitbreakers in SF_6 setups. For simulation of these mechanisms in testing practice, one can therefore also make use of the tubular pieces which are connected by an isolator. Thereby, one makes use of the voltage pulses occurring in practice as test pulses for the isolator or the neighbouring elements [*Boeck*1990].Even the coupling of electromagnetic phenomena through the bushings or by the measuring transformers can be tested in this manner. Maximum overvoltages arise when one side of the test circuit is terminated with a capacitance C and the other side is operated as a shortest possible open-circuited line that is charged to the same voltage but to opposite polarity [*Sun et al.* 1991].

Voltage multiplication too can be temporarily realized using the transmission line type of energy storage device; the setup is in principle so arranged that the potential jumps caused by traveling waves on several lines add up at the test object. In an arrangement with two parallel line devices, suggested by *A.D.Blumlein* in 1941, voltage doubling is obtained. Such a Blumlein generator can be built as a double layer strip conductor, for example, the central electrode of which is charged at one end to U_0 with respect to the two outer ones. If one electrode pair at the beginning of the line is short-circuited, the resulting discharge wave causes a voltage jump of $2.U_0$ at the test object connected between the outer electrodes at the end of the line. Generators of this kind have proved their merit particularly in plasma physics applications. An improvement of this method finally led to the development of "spiral generators" ,which can produce triangular voltage pulses of up to some 100 ns duration; their amplitude is a large multiple of the charging voltage [*Fitch, Howell* 1964].

1.3.9   Limiting Conditions for Impulse Generators

Limiting values that can be achieved in practice shall be estimated for the impulse generators. The maximum steepness or the minimum time to peak of a voltage pulse is limited by the inductance L_d of the circuit (Fig. 1.44).

For example, for a 2 MV impulse generator functioning in air with a specimen of capacitance $C_b = 1$ nF, the minimum time-to-peak works out to about 500 ns ($L_d \geq 20\ \mu H$). Shorter rise-times can be achieved only through additional circuits with sphere-gaps that steepen the rise-time (section 1.3.8). In compact generators (in oil or SF_6), rise-times of about 200 ns at 2 MV can be realized. The maximum time-to-peak is restricted by the utilisation factor to a great extent.

The highest output voltage is about twice the charging voltage U_o. In practice, the utilisation factor depends considerably on the specimen (capacitance) and the damping (R_d).

The longest required time to tail T_h lies around 10 ms. During still longer tail-times, rupture of the current through the switching gaps takes place, which becomes noticeable by voltage collapses on the tail of a switching impulse voltage. In addition, charging of the impulse capacitances must be done via charging switches since otherwise too long charging times would result. Short tail-times result in a low utilisation factor.

1.3.10 Generation of Switching Impulse Voltages with Transformers

To generate switching impulse voltages with times to crest in the millisecond range, besides the usual impulse generators, impulse-excited testing transformers may also be employed. An abrupt rise of the voltage in the excitor winding leads to a transient phenomenon between the transformer and the high-voltage side capacitors. The voltage produced at the latter is utilized as a switching impulse voltage. For impulse excitation, supply is possible from the a.c. mains [*Kind, Salge* 1965], as well as from charged capacitors [*Mosch* 1969]. Both methods have stood the test of practical application [*Anis et al* .1975; *Thione et al.* 1975] and are illustrated by their basic circuit arrangement in Fig. 1.47a. Examples of possible high-voltage impulses are shown in Fig. 1.47c.

The time dependence of the curve up to the peak value $\hat{U}$ takes the form $(1 - \cos \omega t)$, which can be described by the equivalent circuit of a series resonant circuit for this experimental setup (Fig. 1.47b). Here the transformer is replaced by its leakage inductance Ls, the high-voltage side capacitances are represented by C. Assuming a stable voltage source in the excitation winding, for small damping of the circuit, the time to crest is estimated to be:

$$T_p \simeq \pi \sqrt{L_s C}$$

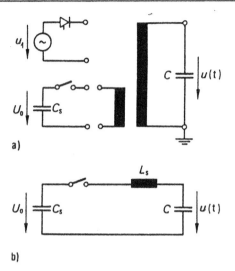

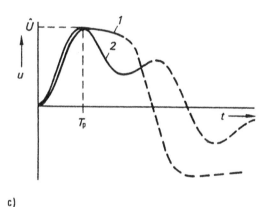

Fig. 1.47 Generation of switching impulse voltages with test transformers
a) basic circuit for excitation from supply network and from capacitance
b) equivalent circuit
c) voltage waveforms with various switching equipment, 1: thyristor, 2 : contactor

The voltage curve after the peak remains more or less constant if a switching rectifier is used in the excitor circuit, e.g. a thyristor. A switch without rectification, e.g. a mechanical switch, produces an oscillating output voltage. The dashed part of the curve shown in the figure is determined by the non-linear magnetization behaviour of the iron core.

This method is especially suitable for the generation of high switching impulse voltages with long times to crest. The impulse form is only marginally adjustable. However, one should take care that transient phenomena within the testing transformer do not lead to overloading [*Wehinger* 1977]. This circuit is often used when for a test two switching impulse voltages are required, e.g. while testing circuit breakers, or while testing the phase to phase insulation. One of the switching impulse voltages is generated with an impulse generator while the other switching impulse voltage is generated with a transformer. Finally, it should be mentioned that high impulse voltages of short duration can also be generated using inductive circuits. For this purpose a high current is passed through the series combination of a high-voltage inductance and a switching device. The test object is connected in parallel with the switching device. If the resistance of the switching device increases strongly and the circuit current is maintained by the action of the inductance, a voltage pulse appears at the terminals of the test object. Exploding wires, for instance, have been found suitable as switching devices [*Salge* 1971].

Measurement of High Impulse Voltages

1.3.11 Peak Value Measurement with a Sphere-Gap

The use of sphere-gaps for the measurement of the peak value of high alternating voltages was described in section 1.1.10. From investigations on the breakdown of gases it is known that the development of a complete breakdown of such a system takes only a few μs at the most, if the applied voltage exceeds the peak value of the breakdown voltage $\hat{U}_d$ for alternating voltages. It follows that sphere-gaps can be used to measure the peak value of impulse voltages, the duration of which is not too short. The limit is approximately $T_2 \geq 40\ \mu s$.

It is assumed here that the air in the space between the spheres contains enough charge carriers to initiate the breakdown without delay after a definite field strength has been reached. By artificial irradiation, using UV sources or radioactive sources, the breakdown region can be sufficiently pre-ionised, so that the statistical scatter of the breakdown time is reduced. The relevant specifications therefore recommend that artificial irradiation be particularly used for the measurement of impulse voltages less than 50kV.). Even irradiation from the arc of a not-encapsulated sphere-gap of an impulse generator can, in the event of a direct line-of-sight connection, be sufficient as a source of irradiation (*Kachler* 1975).

A special feature of measuring the peak value of impulse voltages with sphere gaps is the fact that, on the basis of the occurrence or absence of a breakdown alone, one cannot ascertain how close the peak value $\hat{U}$ of the applied impulse voltage lies to $\hat{U}_d$. This can only be determined by repeated impulses.

To this end, the amplitude of a sequence of impulse voltages is systematically varied until about half the impulses lead to breakdown, i.e., the breakdown probability $P(U)$ is about 50%. For this impulse voltage we then have

$$U_{d-50} \approx \hat{U}_d \approx d \cdot \hat{U}_{d0},$$

where d represents the relative air density and $\hat{U}_{d0}$ is the breakdown voltage under standard conditions; the latter may be obtained from tables and depends upon the sphere diameter, polarity and spacing. The distribution function $P(\hat{U})$ of the breakdown voltage, shown in Fig.1.48, may be determined by repeatedly stressing an electrode arrangement. It can be seen that the withstand voltage U_{d-0} and the assured breakdown voltage U_{d-100}, corresponding to a breakdown probability of 0% and 100% respectively, can only be approximately defined and are therefore not suitable as characteristics.

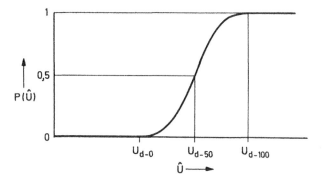

Fig.1.48 Distribution function of breakdown voltage of a sphere-gap for impulse voltage

Instead of the 50% breakdown probability, which can usually only be accurately set for a large number of impulses, one can adjust to a value of $P(\hat{U})$ just below and another just above; the desired value U_{d-50} is then obtained approximately by interpolation. The distribution function of the breakdown voltage is described in detail in Appendix 4.6, along with a special method for the more exact determination of U_{d-50}.

As per the relevant specifications (VDE 0433-2; IEC Publ. 52), sphere-gaps are suitable for the measurement of the peak values of alternating and lightning impulse voltages with a measuring uncertainty of ± 3%. Investigations with switching impulse voltages have shown that the tabulated values in the specifications for lightning impulse voltages can also be used for the measurement of the peak values of switching impulse voltages with a measuring uncertainty of ± 5% (*Gockenbach* 1991; IEC Publ. 60-1). Even during this measurement adequate irradiation must be ensured.

1.3.12 Characteristic Parameters of the Transient Response of Impulse Voltage Dividers[8]

The temporal shape of an impulse voltage is measured using a cathode-ray oscilloscope (KO) or a digital recorder (DR). The quantity to be measured is fed in via a coaxial measuring cable, the input end of which is connected to the secondary terminals of a voltage divider wired to the measuring point (test object). The divider leads, divider, measuring cable and the transient recorder (TR, realized as KO or DR) together constitute the measuring system. If the peak value $\hat{U}$ alone is to be measured, then a direct indicating electronic device (impulse voltmeter) may be also connected.

To investigate the response of measuring systems, test functions are used. Characteristics are conveniently derived from the response to a step function. This method is suitable for theoretical as well as for experimental investigations.

One may consider the measuring system to be generally represented by a four terminal network. A unit step voltage of amplitude $U_{1\infty}$ is applied as the input quantity:

$$u_1(t) = U_{1\infty}\ s(t).$$

The output voltage obtained is:

$$u_2(t) = U_{2\infty}\ g(t)$$

with $U_{2\infty}$ as rated value after the transient oscillations have died down. In this equations $g(t)$ is the normalised step response to the unit step function $s(t)$. In linear systems the voltage $U_{2\infty}$ is proportional to $U_{1\infty}$. The expression $U_{1\infty}/U_{2\infty}$ is called the transformation ratio. An important characteristic for defining the response behaviour of a divider is the response time T, defined by the area:

$$T = \int_0^\infty [1 - g(t)]dt\,.$$

---

[8] Comprehensive treatments in *Schwab* 1969; *Zaengl* 1970; *Hylten-Cavallius* 1970; *Kuffel, Zaengl* 1984; *Beyer et al.* 1986.

Table 1.1

Description	Time domain	Picture domain
Input voltage as step function	$u_1(t) = U_{1\infty}s(t)$	$U_1(p) = \dfrac{U_{1\infty}}{p}$
Output voltage as step response	$u_2(t) = U_{2\infty}g(t)$	$U_2(p) = \dfrac{U_{2\infty}}{p}G(p)$
Transfer function		$G(p) = \dfrac{U_{1\infty}}{U_{2\infty}}\cdot\dfrac{U_2(p)}{U_1(p)}$
Response time	$T = \int_0^\infty [1-g(t)]dt$	$T = \lim_{p\to 0}\dfrac{1}{p}[1-G(p)]$

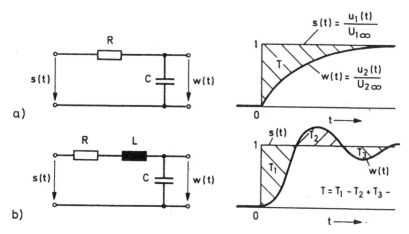

Fig. 1.49 Equivalent circuit and unit set response of impulse voltage measuring systems.
a) RC behaviour, b) RLC behaviour

The simplest case of a four-terminal network with an aperiodic unit step response is shown in Fig.1.49a. This kind of behaviour is named "RC behaviour". Fig. 1.49b shows a four-terminal network with a unit step response incorporating a damped transient oscillation. When the response time is determined here, sub-areas of different sign result. The partial response time T_α can be considered a measure of the reproduction of the front of the step voltage. The overshoot β increases with decreasing damping of the measuring system. The curve of the step response shown is described as "RLC behaviour".

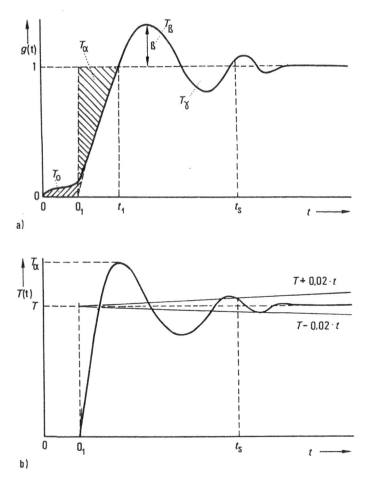

Fig.1.50 Experimental step-response of an impulse voltage measuring system.
a) definitions as per IEC Publ. 60-2
b) evaluation of the step-response time

In practical measuring systems, much more complicated electrical circuits are very often in operation and quite different unit step responses can be encountered. As a result of large overshoot, the response time T can even become negative. For a wide band, and at the same time well-damped measuring system, T_α and β shall be as small as possible.

For theoretical investigations, it is appropriate to apply the Laplace transformation, where the expressions given in Table 1.1 are valid for the parameters introduced.

In pulse technology, quite often the rise-time T_a is used to characterise a step-response. Under this time, one understands the time which the step-response requires to rise from 10% to 90% of its peak value. For an exponential curve,

$$g(t) = 1 - e^{-t/T}$$

the rise time works out to :

$$T_a = 2.2 \cdot T$$

Under the same conditions, for the limiting frequency of the system, we have:

$$f_g = \frac{1}{2\pi T} \cdot$$

As a consequence of the extensiveness of high-voltage measuring circuits, a few difficulties arise in practice during evaluation of the step response, which have led to additional definitions and the introduction of additional characteristics (IEC Publ. 60-2). While sketching, either due to a very slow oscillation or a superposed interference, the beginning of the step response can be determined only inaccurately. But the time T depends strongly on the fixation of the zero point. In such cases, the beginning of the step response O_1 is therefore defined as the point of intersection of a straight extension of the front of the step-response with the zero line(Fig.1.50).

The initial distortion time T_0 affects very strongly the measurement of chopped impulse voltages at the chopping point. A further practical difficulty is the fixation of the static end-value. Here, it must be ensured that the step response is sketched sufficiently long after the settling time t_s and deviates from the value 1 by less than ± 1%. Superposed oscillations e.g., cavity resonances in a laboratory or undamped lead oscillations which lie above 10 MHz are not taken into account.

The settling time t_s is defined as the time from which onwards the remaining response time ($T_r (t)$) of the step response is less than 2% of the settling time t_s (Fig.1.50):

$$T_r(t_s) = \int_{ts}^{\infty} (1 - g(t))\mathrm{d}t \leq 0.02.t_s$$

The requirements on the parameters of the step response of a measuring system for measurement of lightning impulse voltages are summarized in Table 1.2 (IEC Publ.60-2). The overshoot β as a function of the ratio of the partial response time T_α to time to peak T_1 should lie within the hatched area of Fig.1.51.

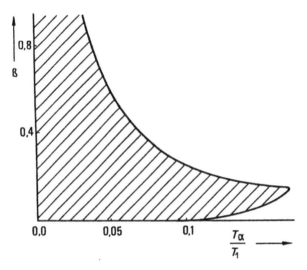

Fig.1.51 Overshoot β as a function of T_α / T_1

Table 1.2

Form of impulse voltage	Parameter			
	$T + T_r (T_c)$	T_α, β	T_0	t_s
Lightning impulse voltages - chopped	–	$\dfrac{T_\alpha}{T_1} = f(\beta)$	–	$< T_1$
Lightning impulse voltages (T_c - Chopping time)	$< 0.05 \cdot T_c$	$\dfrac{T_\alpha}{T_1} = f(\beta)$	$0.005 \cdot T_c$	$< T_c$

1.3.13 Transient Performance of Impulse Voltage Dividers

During measurement of rapidly varying high voltages, the frequency dependence of the velocity of propagation of electrical processes in the usually extended measuring systems can considerably affect the response behaviour. With impulse voltages, appreciable travel time effects occur in the leads, in the divider and in the measuring cable. Transmission in the measuring cable, as a rule, leads to small distortions and can therefore be neglected.

But considerably additional distortions in the step response occur in practice due to a non-optimal construction or earthing of the low-voltage components; these are recognisable in the step response by way of weakly damped superposed oscillations.

For the analysis of an impulse voltage measuring system, voltage divider and lead must be regarded as a whole [*Zaengl* 1970]. Thereby it shall be borne in mind that the elements of the measuring circuit are to be primarily chosen as per the requirements of the test circuit and test voltages to be measured. Detailed requirements on the transient performance of the entire measuring circuit can be adjusted within certain limits by compensation of the low-voltage component of the high-voltage divider (compensating networks).

In the case of not-so-fast phenomena, as during investigation of transient performance with lower frequencies, with switching impulse voltages or while investigating the loading of an impulse voltage generator by a voltage measuring system, the properties of the voltage divider are of decisive significance. The following discussion shall therefore be restricted only to the most important types of voltage dividers. For these types, simple equivalent circuits with lumped components shall be used, which are adequate for working out several problems.

a) Resistive Voltage Divider
In measuring systems with resistive dividers, as in Fig. 1.52a, it is useful for the measuring cable K to be terminated at the TR with its surge impedance Z, thus loading the divider with an effective resistance of the same value. The lead, with its surge impedance, will be terminated either

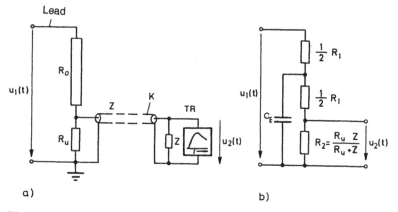

Fig. 1.52 Impulse voltage measuring system with resistive divider
a) circuit diagram, b) equivalent circuit with earth capacitance lead

at the beginning or at the end with a resistance $R_z = Z$ ($\approx 300\ \Omega$). The most important disturbance of the ideal behaviour of the divider is brought about by the earth capacitance of the high-voltage branch, which must necessarily be long, for reasons of insulation at higher voltages. This earth capacitance is taken into consideration, to a first approximation, in the equivalent circuit of 1.52b as C_E, connected at the middle of $R_1 = R_0 + R_z$. Using the relationships introduced under 1.3.12, the unit step response of this circuit can be derived as:

$$g(t) = 1 - e^{-t/T_R}$$

For the time constant, using the approximation $R = R_1 + R_2 \gg R_2$, we have:

$$T_R \approx \frac{1}{4} R.C_E.$$

The output voltage tends to the limiting value

$$U_{2\infty} = U_{1\infty} \frac{R_2}{R_1 + R_2}.$$

g(t) corresponds to the curve shown in Fig. 1.49a, and the desired response time T is equal to the time constant T_R. Assuming a homogeneous distribution of the earth capacitances, it can be shown that C_E is equal to 2/3 of the total earth capacitance C_e effective at R_1. It follows, therefore, approximately:

$$T \approx \frac{1}{6} R.C_e.$$

For vertical cylindrical dividers a value of 12... 20 pF per metre height can be taken for C_e. Hence, e.g., for a 1 MV divider with resistance $R = 10\ k\Omega$ and 3 m high, the earth capacitance is 45 pF and thus the response time $T = 75$ ns.

Resistive dividers are conveniently used for the measurement of steep impulse voltages of not too long a duration. Dividers for switching impulse voltages must be built with a large resistance R because of the heating and loading of the voltage source, which results in an unfavourable transient response for rapid voltage variations. By optimal field control, the effect of the earth capacitance can be appreciably reduced such that with even with higher resistances sufficiently good transfer properties could be attained [*Peier, Stolle* 1987]. For voltages above 1 MV, the practical construction of fast response resistive dividers becomes increasingly difficult, since one must try to compensate for the effect of the earth capacitances by increasing the coupling capacitances to the high-voltage electrode. One then has a capacitively controlled resistive divider which does however have a considerable capacitance parallel to the divider resistance. Moreover, this capacitance can be caused to oscillate with the inductance of the measuring circuit; in this way the system acquires RLC behaviour.

In order that a resistive voltage divider shows RC behaviour, the lead must be terminated with the surge impedance of the lead, of about 300 Ω, either at the beginning or at the end. If this is not the case, the transient performance shows an oscillation which is dependent on the length of the lead.

b) Capacitive Voltage Divider

In measuring systems with capacitive dividers, as in Fig. 1.53a, the measuring cable K cannot usually be terminated at the TR, since C_2 would discharge too rapidly because of the usual order of magnitude of the surge impedance ($Z \approx 75\ \Omega$). The series matching with Z indicated in the figure has the effect that only half the voltage at the divider tap enters the cable, yet this is doubled again at the open end, so that the full voltage will be measured at the TR once more. On the other hand, the reflected wave may find matching at the cable input, since for very high frequencies C_2 acts as a short-circuit. The transformation ratio therefore changes from the value

$$\frac{C + C_2}{C}$$

for very high frequencies, to the value

$$\frac{C + C_2 + C_K}{C}$$

for lower frequencies. However, very often the capacitance of the measuring cable C_K can be neglected in comparison to C_2.

Whilst the earth capacitance C_e of capacitive dividers can be taken into account by a correction of the divider ratio (see also 1.1.11), the response behaviour here is essentially determined by the inductance of the divider lead and of the divider itself. As a first step, in the equivalent circuit of Fig. 1.53c an inductance L has been assumed in series with C. We obtain for this circuit:

$$g(t) = 1 - \cos\ \omega t \quad \text{with } \omega^2 = \frac{1}{L}\frac{C + C_2}{C.C_2} \approx \frac{1}{LC}$$

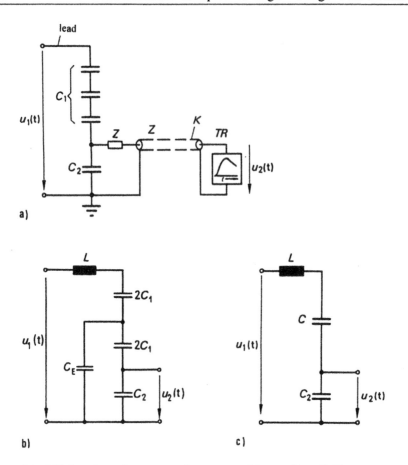

Fig. 1.53 Impulse voltage measuring system with capacitive divider
a) circuit diagram, b) equivalent circuit with lead inductance and earth capacitance,
c) simplified equivalent circuit

since usually $C_2 \gg C$. The output voltage strives to attain the limiting value

$$U_{2\infty} = U_{1\infty} \frac{C_2}{C} \; .$$

A measuring circuit with a capacitive divider indicates in practice a basic oscillation (RLC behaviour as per Fig. 1.49b), which is primarily determined by the circuit data L and C and damped by the losses in the capacitors and the ohmic losses in the leads. An overshoot β upto 80% is common. In addition, high-frequency oscillations which have their origin in the low-voltage part or high-voltage part of the voltage divider get superposed. The capacitance C is so dimensioned that the effect of variation of the earth capacitance C_e on the transformation ratio would be negligible (C>40 pF/m).

Capacitive dividers can be advantageously applied for measuring switching impulse voltages. Measurement of lightning impulse voltages is possible only in small impulse voltage circuits since at higher voltages oscillations are induced in the measuring circuit. The frequency of the oscillations increases with reducing capacitance C. Resonance frequencies above a few MHz are no longer excited by lightning impulse voltages. The capacitance of the voltage divider increases the front-time of the impulse voltage and must therefore be included in the load capacitance.

c) Damped Capacitive Voltage Divider

If a resistance is included in series with the capacitors, one obtains the damped capacitive divider (Fig. 1.54). Dimensioning of the high-voltage capacitance follows on the same lines as in a capacitive divider. The dimensioning of the resistance R_1 and its arrangement determines the type and transfer performance of damped capacitive voltage dividers. Dimensioning of the resistances according to the aperiodic limiting case of a transmission line chain [*Zaengl 1964, Beyer et al.* 1986] results in an optimal damping value of $R_1 \approx 4\sqrt{L/C_e}$. In practice, a constant resistance value of above 1000 Ω results, independent of the height of the voltage divider ($L \approx 1\mu H/m$, $C_e \approx 12$ pF/m). For an optimal transfer performance, a damped capacitive voltage divider so dimensioned must have as termination a damping resistance at the beginning of the lead. Its transfer performance is then similar to that of a low-ohmic resistive voltage divider.

The practical difficulties with a damping resistance at the beginning of the lead at high voltages and the pulse distortion due to the voltage divider under low capacitive loads [*Feser* 1974] lead to a dimensioning of the resistance R_1 according to the aperiodic limiting resistance of the measuring circuit.

$$R_1 \approx (1...2).\sqrt{L/C}\ \cdot$$

Such a dimensioning results in resistances of ca. 200 ... 400 Ω. With that, even the lead is approximately terminated. In small voltage dividers ($h < 1$ m), the resistance R_1 is also arranged as a concentrated one at the top of the voltage divider.

A damped capacitive divider behaves for high frequencies like a resistive divider and for low frequencies like a capacitive divider. The time constant

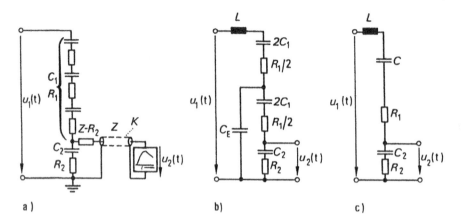

Fig. 1.54 Impulse voltage measuring system with damped capacitive divider
a) circuit diagram, b) equivalent circuit with lead inductance and earth capacitance,
c) simplified equivalent circuit

of the low-voltage portion $R_2.C_2$ is matched with the time constant $R_1.C$ of the high-voltage portion. Transfer performances of damped capacitive dividers so dimensioned indicate an RLC behaviour (Fig. 1.49b), whereby the overshoot β, depending on the choice of R_1, can approach the optimal value of about 6%. The response time is therefore smaller than that of a corresponding low-ohmic resistive divider $\left(T < \dfrac{1}{6}R_1.C_e \right)$. The effect of earth capacitance on the transformation ratio is the same as in the capacitive divider. The damped capacitive divider can therefore be utilised in a wide frequency range, i.e. for impulse voltages with very different durations and also for alternating voltages.

For calculation of the transmission properties of complete measuring systems, the voltage dividers are usually simulated as homogeneous transmission line networks and the lead, as a loss-less line [*Zaengl* 1970]. If the damped capacitive voltage divider is combined with high-ohmic parallel resistances, one obtains a divider suitable for measurement of direct voltages, alternating voltages and impulse voltages.

1.3.14 Experimental Determination of the Response Characteristics of Impulse Voltage Measuring Circuits

For exact measurement of rapidly varying voltages the complete impulse voltage measuring circuit indicated schematically in Fig. 1.55 must be

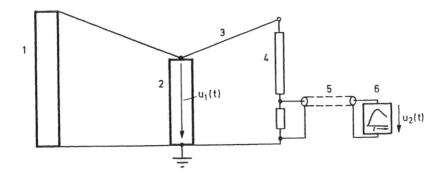

Fig. 1.55 Circuit of a complete impulse voltage measuring system
1 Impulse voltage generator
2 Test object
3 Divider lead
4 Divider
5 Measuring cable
6 TR

considered. Here the voltage $u_1(t)$ to be measured is the voltage at the terminals of the test object, whilst the measured value $u_2(t)$ corresponds to the curve measured with TR. The response time of the entire measuring system, T_{res}, is obtained from the response time T of the divider with lead, the response time of the coaxial measuring cable T_K and the response time of the transient recorder T_{TR}. If all three components show RC behaviour, to be verified in the case of the divider, then the resultant response time of the system can be calculated from the following equation:

$$T_{res} = T + T_K + T_{TR} \cdot$$

The effect of the individual components on the resultant response time can be estimated from this relationship. T_K is usually much smaller than T, and can be neglected in the case of high-quality cables which are not too long. Impulse voltage oscilloscopes or transient recorders normally have a limiting frequency of over 50 MHz, from which $T_{TR} \leq 3$ ns can be calculated. In most cases the response behaviour of the divider determines the response time of the entire system, so that one may take $T_{res} \approx T$.

Investigations to determine the step response of a system can be carried out using high-voltage as well as low-voltage. In the first case one makes use of the voltage collapse of a spark gap with as homogeneous a field as possible, preferably operating at higher field strengths (compressed air, SF_6, oil) to steepen the collapse. For measurements with a low-voltage square wave generator the signal, reduced by the divider in the divider ratio, requires a sensitive amplifier. One should therefore make sure that the latter's display behaviour does not differ appreciably from that of the TR used for high voltage measurements, or that the difference is suitably taken into account.

Measurement of the transfer performance of an impulse voltage measuring circuit should be done on an experimental setup(lead, electrodes, mounting) that corresponds as far as possible to the later test setup. The response time can then be optimised still with the low-voltage part of the voltage divider [*Feser* 1983].

Another method of determining the response time is by recording the impulse voltage-time curve of a known electrode configuration, using the measuring system to be investigated. The response time can be evaluated from a comparison of the measured impulse voltage-time curve, for linearly rising impulse voltages of constant rate of rise S, with the " true" impulse voltage-time curve of this same configuration. This method of response time determination is based upon the fact that the steepness of a wedge-shaped impulse voltage, after a certain transient and damping period is reproduced accurately by all measuring systems in question. This behaviour is shown in Fig.1.56. For a known transfer function $G(p)$,for example, the curve can be calculated using the relationship

$$U_2(p) = \frac{U_{2\infty}}{U_{1\infty}} U_1(p)G(p).$$

For a divider corresponding to the equivalent circuit of Fig.1.52b, with $u_1(t) = S.t$, we find:

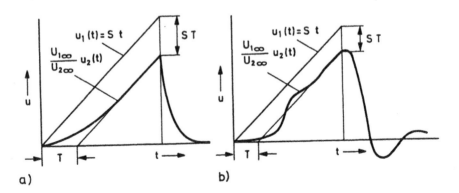

a)   b)

Fig. 1.56 Display of a wedge-shaped impulse voltage
a) system showing RC behaviour b) system showing RLC behaviour

$$u_2(t) = \frac{R_2}{R_1 + R_2} S[t - T(1 - e^{-t/T})].$$

If the steepness is accurately reproduced, the response time T can be determined from the voltage error ST. In practice, several measurements are made with as many different values of S as possible, and the response times obtained are averaged. This method has the advantage that the measuring system can be investigated by a test method coming close to the actual requirements and at comparatively high voltages. Fundamental investigations of electrode configurations in air have shown that, for sphere-gaps with only a weak inhomogeneous field, the true impulse voltage-time characteristic referred to standard conditions can be approximated by:

$$U_d = \hat{U}_{d0} + \sqrt{2FS}.$$

Here $\hat{U}_{d0}$ is the static breakdown voltage according to section 1.1.10 and F the voltage-time area of the gap as discussed in 3.8.1b. Since the voltage-time areas of geometrically similar configurations are, to a good approximation, proportional to the static breakdown voltage, the above relationship allows easy conversion from known impulse voltage time characteristics to other configurations. As a test gap the IEC has suggested a single pole earthed sphere-gap with $D = 250$ mm and $s = 60$ mm, and a static breakdown voltage $\hat{U}_{d0} = 161$kV for standard conditions and negative polarity; the impulse voltage-time characteristic of this gap has been determined by international comparison measurements. These measured values are obtained from the above equation with sufficient accuracy, if the value 2 kV.µs is substituted for F. For an other sphere gap with a static breakdown voltage of $\hat{U}*_{d0}$ the voltage-time area works out to:

$$F^* = F \frac{\hat{U}^*_{d0}}{\hat{U}_{d0}}.$$

1.3.15 Calibration of Impulse Voltage Measuring Systems

Impulse voltage measuring systems must be regularly calibrated and checked. The suitability of a measuring system for a particular measuring assignment can be ensured by two separate methods [Kind et al. 1989]. The comparison method compares the forms of output voltages of the measuring system to be calibrated to those of a reference system which has been calibrated for this assignment by an accredited laboratory. The response parameter method measures the transformation ratio of the voltage divider and the response time and evaluates the parameters of the step response (IEC Publ. 60-2). The transformation ratio of voltage dividers can be determined by measuring the impedances, by measuring the input and output voltages or with a gap (e.g. sphere-gap).

1.3.16 Feedback-Free Voltage Measurement with Field Sensors

Due to their spatial dimensions, e.g. at 2 MV, high-voltage dividers have a bandwidth of a few MHz. Yet, in order to capture the front oscillations and oscillations at the peak, sensors, which measure the electric field can be applied to sketch the pulse form.(IEC Publ. 60-2). The capacitive field sensors are to be calibrated at a specific place for voltage measurement. In case discharges occur on the field sensor or charges flow-in, the measurement would be in error. The field sensor should be spatially closely coupled with the voltage to be measured. This means that it is advantageous to make use of a potential-free, optical transmission of the measured signals. The bandwidth of sensors is dependent on their size, but bandwidths of a few 10 MHz are easily achieved [*Feser* 1984]. In high-voltage testing practice, spherical sensors, which also capture the direction of the field, process the measured values within the sphere and provide optical transmission to the ground level, have proved successful [*Feser, Pfaff* 1984]. In encapsulated SF_6-insulated systems, as a rule, a plane field sensor is built in on the earth side as a voltage measuring device. Transmission of the measured value can then be made by means of a measuring cable.

1.3.17 Measuring Instruments Associated with High-Voltage Dividers

For measuring the peak value of impulse voltages, impulse voltmeters are used (IEC Publ. 790). The accuracy of measurement of the peak value depends upon the form of the impulse voltage to be measured. Fundamental difficulties arise due to the one-time nature of the phenomenon, the short time for which the peak value is present and the long storage time required to enable a reading. The principle of peak value measurement is shown in Fig. 1.57.

   The classic (passive) diode rectification with storage in C_m requires a small time constant $R_i \cdot C_m$ and a large discharge time constant $R_m \cdot C_m$,

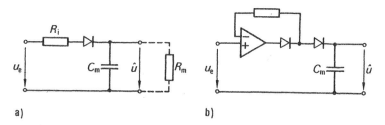

Fig. 1.57 Principle of measurement of peak value of impulse voltages
a) with diode rectification, b) with operation amplifier (e.g. impedance transformer, $I_e < 10^{-12}$ A)

whereby R_m could be the input resistance of an analogue or digital indicating instrument. The difficulties lie in the non-ideal properties of the diode. The commonly used circuit today, therefore, is that with operation amplifiers(Fig. 1.57b) built for 1 or 2 stage operation [*Beyer et al.* 1986].

For measuring the impulse waveform, special analogue impulse oscilloscopes or digital recorders are available as transient recorders which must satisfy special conditions (IEC Publ. 790, IEC Publ. 1083-1). In contrast to the usual commercial oscilloscopes, the impulse oscilloscopes have no y-amplifier i.e., they are built for small vertical deflection sensitivity. The most important differences are thus between a low sensitivity against interference voltages and a high accuracy and stability of the deflection system. The bandwidth of these impulse oscilloscopes lies above 30 MHz.

The central piece of a digital recorder is the analogue/digital converter. The time dependent analogue impulse is continuously sampled,, quantised and stored. The storage depth determines the number of samplings or the duration (length) of the recording. The sampling time gives the time resolution and the quantising determines the vertical resolution. For impulse voltage measurements, usually digital recorders with a vertical resolution of at least 8 bit (0.4% amplitude resolution), a sampling frequency of more than 20 MHz and a storage depth of 2 K (2048 samplings) are necessary. The commercially available digital recorders must be specially screened against the high interference voltages in the high-voltage testing area. Particular attention must be devoted to the short-time and long-time stability of these instruments. If the digital recorder is to be applied as a measuring equipment, it must fulfill various conditions and must be calibrated regularly (IEC Publ. 1083-1). The advantage of the digital recorder is in the convenient further processing of the measured values.

1.4 Generation and Measurement of Impulse Currents

Rapidly varying transient currents of large amplitude, as a rule, appear in connection with high voltages, namely through the discharge of energy storing devices. They often develop as a consequence of breakdown discharge mechanisms and are frequently accompanied by large forces and high temperatures.

If these currents have a definite shape, they are referred to as impulse currents; among other things, these are required for the simulation of lightning and short-circuit currents during tests on service equipment. Examples of the specific application of the physical effects of impulse currents are magnetic field coils for the confinement of plasmas, electrodynamic drives or gaps as impulse radiation sources.

The measurement of rapidly varying high currents is usually performed with measuring resistors, or with arrangements which exploit the inductive effect of the current to be measured.

1.4.1 Characteristic Parameters of Impulse Currents

Impulse currents can have very different shapes, depending upon their application and occurrence. Quite often impulse currents appear as aperiodic or damped oscillatory currents, and as alternating currents with a duration of only a few half-periods. The maximum instantaneous value of the current is denoted the peak value $\hat{I}$; characteristic parameters for the time dependence will be mentioned here only for impulse currents intended for testing.

To simulate currents produced by lightning strokes, single, unidirectional impulse currents of short duration are used, which reach a peak value $\hat{I}$ rapidly without appreciable oscillations and then decrease to zero. The characteristics of these double exponential impulse currents are defined in Fig. 1.58a (VDE 0432-2; IEC Publ. 60-1). Usual values are $T_1 = 4$ μs or 8 μs and $T_2 = 10$ μs or 20 μs. For metal-oxide arrestor, additionally $T_1 = 1$ μs or 30 μs and $T_2 = 10$ μs or 80 μs are to be used (VDE 0675). The peak value $\hat{I}$ of the current during high-current testing = 100 kA ($T_1/T_2 = 4/10$) and during the residual voltage test = 20 kA($T_1/T_2 = 8/20$). The undershoot may not be more than a maximum of 20% of $\hat{I}$.

Besides the peak value $\hat{I}$ and the impulse form (T_1/T_2), very often in testing practice, the charge ($\int i dt$) the quadratic integral ($\int i^2 dt$) and the steepness (di/dt) of a lightning discharge, usually with separate test circuits,

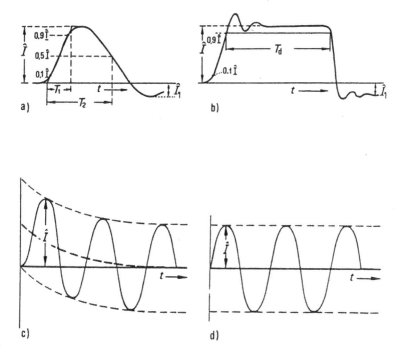

Fig. 1.58 Examples of impulse currents
a) double exponential impulse current, b) rectangular impulse current, c) sinusoidal impulse current with exponential d.c. component, d) sinusoidal impulse current without d.c. component

are made use of for simulating the lightning discharges. Rectangular impulse currents appear during the discharge of long transmission lines. The duration of these impulses is the time T_d, during which the current remains greater than 0.9 $\hat{I}$ (Fig. 1.58b). For the testing of overvoltage arrestors, therefore, even rectangular impulse currents whose duration and amplitude correspond to the rated voltage of the supply network, are used (VDE 0675). Impulse currents which occur in a.c. networks during short-circuits are alternating currents which may be superimposed on an exponentially decaying d.c. component. Here the highest instantaneous value of the current determines the dynamic stress to which the components of the setup are subjected; it is named as the impulse short-circuit current. For short-circuit tests on operating equipment a current shape as in Fig. 1.58c is aimed at, which represents a particularly high stress (VDE 0670-1). For an appropriate switching instant, a current can also appear without the d.c. component, as shown in Fig. 1.58d.

Generation of Impulse Currents[9]

1.4.2   Energy Storage Systems

For the generation of high impulse currents the power which may be drawn from the power supply network is not normally sufficient to obtain a current of given shape and amplitude. In these cases one has to resort to energy storage systems which can be discharged with much greater power than is required to charge them. In principle, capacitors, inductors, transmission line type storage devices, rotating machines, accumulator batteries and even explosives are available as energy storage devices. The use of batteries and explosives is restricted to special cases only; these will not be discussed here further.

a) Capacitive Energy Storage
The energy stored in a capacitor of capacitance C at the voltage U_o is given by

$$W = \frac{1}{2}CU_0^2 .$$

It follows that the energy density in the dielectric stressed at the field strength E is given by

$$W' = \frac{1}{2}\varepsilon_0\varepsilon_r\, E^2 .$$

If one substitutes $\varepsilon_r = 4$, $E = 1000$ kV/cm, the values achievable with oil-impregnated paper, one obtained $W' = 0.2$ Ws/cm^3. Capacitors are energy storage devices of high quality and extremely suitable for power amplification. They are able to store energy over a long period of time. The time constant for discharge through its own insulation resistance often reaches the order of hours. Consequently, a capacitive energy storage device can be charged by a source of low power.

The largest storage systems of this type were built for experimental investigations in plasma physics to generate high magnetic fields; their energy content is a few MWs and the charging voltage some tens of kV. When these systems discharge, currents of several tens of MA are obtained. Further fields of application are test setups for surge diverters, lightning

---

[9] Comprehensive treatment, among others, in *Craggs, Meek* 1954; *Sirotinski* 1956; *Früngel* 1965; *Knoepfel* 1970; *Beyer et al.* 1986.

current simulation and electro-hydraulic metal forming. Diverse installations and their working principles are described in the literature [*Mürtz* 1964; *Prinz* 1965; *Bertele, Mitterauer* 1970 *Modrusan* 1976].

b) Inductive Energy Storage
The energy stored in a coil of inductance L for a current I_0 is given by

$$W = \frac{1}{2} L I_0^2 \, .$$

For the energy density in the space filled with magnetic flux of density B, we have :

$$W' = \frac{1}{2} \cdot \frac{B^2}{\mu_0 \mu_r} \, .$$

Inductive energy storage systems are built with air-cored coils, since the maximum value of B is restricted to about 2T in magnetic materials by saturation process. The maximum value of energy density is limited by the heating of the leads and by the magnetic forces. A value of $B = 10$T can readily be achieved with coils under normal conductivity conditions, so that $W' = 50$ Ws/cm^3. This value lies well above the energy density attainable in the electric field of a capacitor.

A major limitation in the application of inductive energy storage devices with conductors under normal conductivity conditions is the self-discharge time constant L/R, which is of the order of a few seconds and is very low compared with a capacitor system. Normally conducting storage coils therefore have to be charged with high power and can only be used as short-term storage. Long-term inductive storage may only be realised using superconducting coils.

c) Mechanical Energy Storage
In mechanical energy storage devices the energy is stored in moving masses and can be released by abrupt deceleration. The kinetic energy stored in a mass m moving with velocity v is:

$$W = \frac{1}{2} m v^2 .$$

In gyrating masses, energy densities greater even than those of the magnetic field can be realized. The maximum value of the energy density is limited by the centrifugal forces. For a steel flywheel with a peripheral velocity of 150 m/s, a value of $W' = 100$ Ws/cm^3 results.

As a rule, the rotors of suitably designed generators function as gyrating masses for mechanical energy storage devises; the conversion of their kinetic energy into electrical energy is achieved by deceleration. Generators of sinusoidal impulse currents are designed as synchronous generators and those for generating unipolar impulse currents usually as unipolar generators.

As a result of their high densities, mechanical energy storage systems can be built with capacities up to the order of 1000 MWs and owing to the very large self-discharge time constant, low-power charging equipment is quite adequate.

1.4.3   Discharge Circuits for the Generation of Impulse Currents

The aim of impulse current circuits is to generate a rapidly varying transient current of specified form and amplitude in a given arrangement. This may be needed to test the withstand capacity of operating equipment against stress by an impulse current, or to repeatedly trigger certain physical effects such as the excitation of magnetizing coils. Analogous to impulse voltage circuits, the arrangement in which a given impulse current is to be produced shall be designated the test object.

a) Circuits with Capacitive Energy Storage
The equivalent circuit of an impulse current circuit with capacitive energy storage is shown in Fig. 1.59a. L_1 and R_1 represent the unavoidable inductance and ohmic resistance respectively of the impulse current circuit.

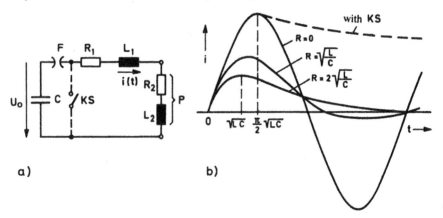

Fig. 1.59 Impulse current circuit with capacitive energy storage
a) equivalent circuit, b) current curves

If the test object P consists of a resistance R_2 and an inductance L_2 in series, and if one groups $L_1 + L_2 = L$ and $R_1 + R_2 = R$ then the current curves shown in Fig. 1.59b are obtained for different values of resistances R on ignition of the three-electrode gap F, most commonly used as a switch [*Deutsch* 1964; *Bertele, Mitterauer* 1970].

In this kind discharge circuit the impulse current is critically influenced by the test object. The time variant form of the impulse current and the characteristic parameters of lightning discharges are calculated for this circuit in Appendix 4.4. The highest peak value of the current is reached for the case of low damping, i.e., when

$$R \ll 2\sqrt{\frac{L}{C}}.$$

In this case, with the capacitor charged to a voltage U_0, we have for the peak value of the discharge current:

$$\hat{I} \approx U_0\sqrt{\frac{C}{L}} = \sqrt{\frac{2W}{L}}.$$

The maximum rate of rise of the current at $t = 0$ is :

$$\left(\frac{di}{dt}\right)_{max} \approx \frac{U_0}{L}.$$

To increase the value of I, and for maximum rate of rise of the current, one therefore tries to keep L low. This calls for compact assembly of the setup and, if necessary, parallel connection of several capacitor units. If the first current oscillation shall not take less than a prescribed duration, then besides decreasing L the value of C and hence W must be increased at the same time, so that the discharge frequency

$$f = \frac{1}{2\pi\sqrt{LC}}$$

does not become too high. In discharge circuits with low damping, an aperiodic current can be achieved by the introduction of short-circuiters (KS in Fig. 1.59a) [*Bertele, Mitterauer* 1970]. A short-circuit at the current maximum signifies that no voltage exists at the short-circuiting switch KS. An increased expense is necessary for such short-circuiting. After the short-circuit, the current decreases with the time constant L/R, i.e., very high charge flows through the specimen. While short-circuiting at current zero,

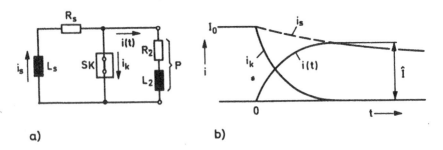

Fig. 1.60 Impulse current circuit with inductive energy storage
a) equivalent current, b) current curves

maximum voltage appears across the short-circuiting switch so that in such a case short-circuiting can be done with a simple sphere gap.

b) Circuits with Inductive Energy Storage

Fig. 1.60 shows the equivalent circuit of an impulse current circuit with inductive energy storage and the most important current curves. Here too the test object P consists of an inductance L_2 and a resistor R_2 in series. With the commutating switch SK closed, the storage coil L_s is charged to a current I_0 (from a source not shown in the figure) via the loss resistance R_s of the whole charging circuit. At $t = 0$ the switch SK is opened; the desired commutation of the switching current to the test object occurs when SK generates a voltage which is sufficiently high and tuned to the test object.

Neglecting the resistances R_s and R_2, from the requirement that the total magnetic flux must be the same before and after commutation, the condition

$$I_0 L_s = \hat{I}(L_s + L_2)$$

may be derived. Then for $L_s \gg L_2$ the full current amplitude can be expected in the load.

The realization of the switching device SK is a special technical problem of inductive energy storage circuits. In addition to arc switches, exploding wires or foils have proved to be good [*Salge et al*, 1970]

c) Circuits with Mechanical Energy Storage and Mains Supplied Setups

Impulse current arrangements with mechanical energy storage are mainly set up when very high energies are required for durations of up to one second. Mains supplied setups are also, in principle, mechanical energy

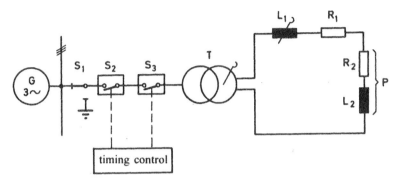

Fig. 1.61 Overall circuit diagram of a mains supplied impulse current setup

storage systems because the energy is initially taken from the kinetic energy of the gyrating masses of the machines in the supply network.

Particularly high power and energies are required for testing high-voltage circuit breakers. Thus, very elaborate arrangements must be set up, often containing short-circuit generators to produce the impulse current [*Slamecka* 1966]. The requirements are usually much less stringent for basic investigations, for instance, on arcs and contacts. In these cases, by connection to a three phase medium voltage network, power of some MVA can be generated in comparatively less complicated setups, as will be shown in the example of Fig. 1.61.

An impulse current transformer T is connected to the network on the primary side via the earthing switch S_1, the series connection of the safety circuit breaker S_2 and the on-load circuit breaker S_3. The secondary side of the transformer is connected to the test object P via a coil L_1. Initially, after closing S_1, S_2 is closed. Shortly after S_3 is given its "on" command, S_2 can be given its "off" command again, so that both switches will only be in the "on" condition simultaneously for a short period. In this way it is possible to switch-on individual half-cycles of the mains voltage using commercial circuit breakers. The short switching time has the great advantage that the supply network is under load only for short periods; appreciably more power can thus be drawn than would be possible for longer switching times.

d) Impulse Current Circuit with Transmission Line Type Energy Storage
Instead of concentrated capacitors and inductors, a transmission line, for example, a cable can also be used as energy storage for the generation of impulse currents. This is of some practical significance, particularly when

nearly rectangular impulse currents are to be generated. The energy densities which can be achieved are the same as those given for capacitive energy storage systems. The current realized during the discharge, for the case of a short-circuit of the lines, is determined by the surge impedance acting as the internal resistance of the system.

In practical applications, a multi-meshed interactive network is usually set up instead of a homogeneous transmission line(Fig.1.62).The rectangular impulse duration can then be calculated from [*Modrusan* 1977] :

$$T_d \approx 2.\frac{n-1}{n}\sqrt{LC} \quad \text{with } L = n.L' \text{ and } C = n.C' .$$

The number of sections n shall be ≥ 8 in order to achieve adequate approximation to the rectangular form. For maximum amplitude of the current, one obtains:

$$\hat{I} = \frac{U_0}{R_a + \sqrt{\dfrac{L}{C}}}$$

with the terminating resistance $R_a = R_1 + R_2$. If the undershoot shall be as small as possible, R_1 must be so chosen that

$$R_a = R_1 + R_2 = \sqrt{\frac{L}{C}} \quad .$$

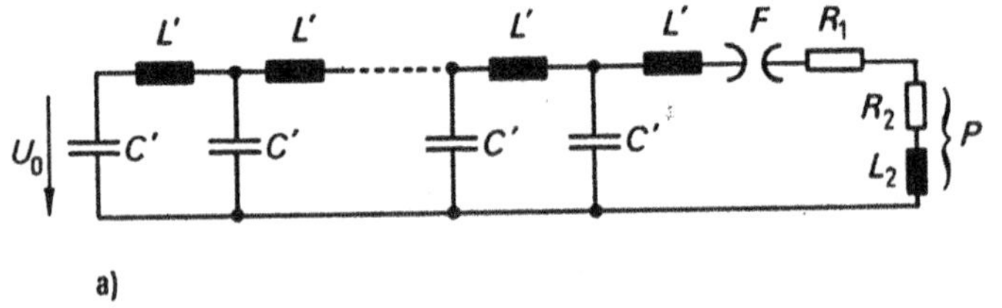

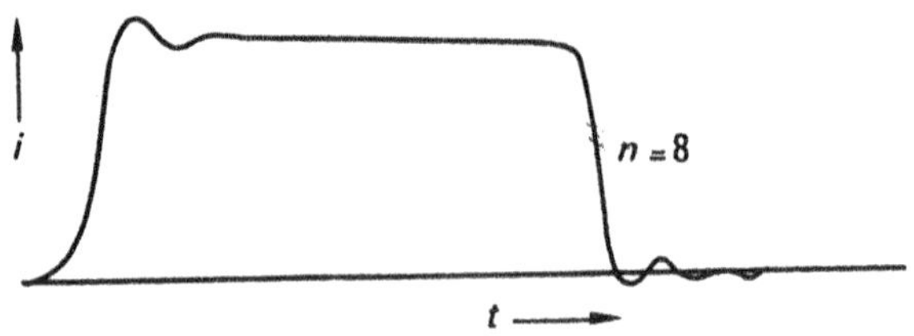

a)

b)

Fig. 1.62 Impulse current circuit with transmission line type energy storage
a) equivalent circuit, b) current waveform

In order that the oscillations at the peak shall, as far as possible, be negligible, C' and L' must not be chosen to be equal.

Measurement of Rapidly Varying Transient Currents

1.4.4 Current Measurement with Measuring Resistors

A current measuring system with a measuring resistor (shunt) is shown in Fig. 1.63. The voltage across the resistor R due to the current $i(t)$ to be measured is fed to the transient recorder TR via a measuring cable. Termination of the cable with the surge impedance Z hardly affects the measuring voltage, if the condition $R \ll Z$ is satisfied. However, magnetic fields caused by the current to be measured and stray magnetic fields can induce voltages in the measuring circuit which are superimposed on the desired measuring signal iR. The following relationship is valid :

$$u(t) = iR + L\frac{\mathrm{d}i}{\mathrm{d}t} + \frac{\mathrm{d}\Phi}{\mathrm{d}t} \;.$$

Induced voltages due to stray magnetic fields Φ can usually be kept low by careful screening. An arrangement of the measuring circuit with low self-inductance L requires the current path within the screening to be such that minimum magnetic flux is linked by the loop formed at the measuring tap. Such a coaxial measuring resistor shows an RC-transfer behaviour as per Fig. 1.49a. The response time works out to

$$T = \frac{1}{6} \cdot \frac{\mu_o}{\rho}\delta^2,$$

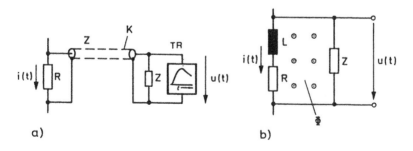

Fig. 1.63 Current measuring system with measuring resistor
a) circuit diagram, b) equivalent circuit

with the wall thickness of the resistance material δ and the specific resistance ρ. The response time increases quadratically with the wall thickness. To control the heating during lightning impulse stressing, one needs mass, i.e., large wall thickness. In practice, one solves both these contradicting demands for a measuring resistor with good transfer properties by slits in the resistance material for small wall thicknesses in order to achieve magnetic field coupling within the resistance cylinder. In large wall thicknesses of the resistance material (a few cm), even the measuring tap is arranged asymmetrically inside the resistance material so as to improve the transfer properties [*Malewski et al.* 1981].

Measuring shunts for impulse currents of short duration can be built with rise-times of a few ns order of magnitude. An example is shown in Fig. 3.101. The resistance element itself, depending upon the value of R, can be made of parallel carbon film resistors or low inductance wire resistors, of parallel resistance wires or of resistance foils. Other modes of construction are described in the literature [*Sirotinski* 1956; *Schwab* 1969; *Gontscharenko et al.* 1966; *Beyer et al.* 1986].

The potential link between the TR and the impulse current circuit, necessary when measuring shunts are used, is sometimes disturbing during high-voltage measurements. It can often lead to the formation of earth loops and thus interfere in the measurements of very fast phenomena.

1.4.5   Current Measurement Using Induction Effects

If two circuits are magnetically coupled, as shown in Fig. 1.64a, the following expression is true:

$$u_2 = -i_2 R_2 - L_2 \frac{di_2}{dt} + M \frac{di_1}{dt}$$

where R_2 and L_2 are the effective resistance and the self-inductance respectively, measurable at the terminals of circuit 2; M is the mutual inductance between circuits 1 and 2.

The first way of measuring i_1 with this arrangement refers the current measurement back to the measurement of u_2. A measuring system with a high internal resistance is connected to the terminals of circuit 2 and with $i_2 = 0$, we have :

$$u_2 = M \frac{di_1}{dt}$$

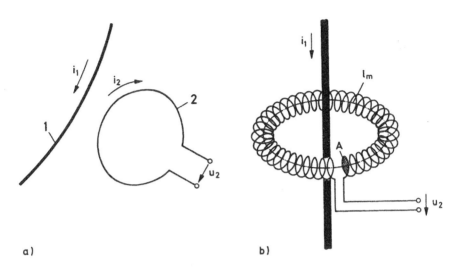

Fig. 1.64 Arrangement of two magnetically coupled current circuits
a) as a conductor loop, b) as a Rogowski coil

In the other arrangement of the measuring loop, often in the form of the Rogowski coil, coil 2 encircles the conductor through which the current to be measured flows, in the manner shown in Fig. 1.64b. From the law of induction it follows directly that, for a uniformly wound coil with N turns, winding area A and length l_m,

$$M = \frac{\mu_0 NA}{l_m}.$$

In order to obtain a parameter which is proportional to the current to be measured, the measured voltage u_2 must be integrated. This can be done most easily by a RC circuit, but more elegantly by using an appropriately wired operational amplifier. To avoid measuring errors as a result of stray magnetic fields, the winding is usually of the criss-cross type.

The differential equation valid for Fig. 1.64a provides further means of current measurement if circuit 2, with the lowest possible effective resistance, is short-circuited. For $u_2 = 0$ and $R_2 = 0$ we have:

$$L_2 \frac{di_2}{dt} = M \frac{di_1}{dt}.$$

If the two circuits are magnetically closely coupled, $L_2 \approx M$ and

$$i_2 \approx i_1 \ .$$

This is a current transformer which is often built with a variable number of turns, i.e., with a transformation ratio differing from 1, and has an iron core for close magnetic coupling. Iron core current transformers are not suitable for measuring rapidly varying currents of short duration, but can be used with great accuracy for phenomena in the ms range and can be designed for high currents.

1.4.6   Other methods of Measuring Rapidly Varying Transient Currents

As further means of measuring impulse currents, one may mention those methods which make use of magnetic field dependent material properties. The Hall generator and magneto-optic elements belong to this category [*Schwab* 1969]. In the Hall generator, through which a constant control current flows and which is permeated by the magnetic field of the current to be measured, the Hall voltage is directly proportional to the measured current. This method became significant after the development of semiconductors with sufficiently large Hall constants. Magneto-optic methods of current measurement use the rotation of the plane of polarization in materials by the magnetic field which is proportional to the current (Faraday effect). Thereby, the material is irradiated with linearly polarized light. Both these methods are insensitive to overloading. Even photo-luminescent diodes, whose momentary light emission is proportional to the current flowing through them, can be used for current measurement [*Wiesinger* 1969]. The advantage of these methods is that they allow galvanic isolation of the measuring setup from the main current circuit.

1.5   Non-Destructive High-Voltage Tests

When an insulation system is investigated, its breakdown voltage defines the upper limit of the voltage range. However, it is usually not possible to draw conclusions about the cause of breakdown discharge from a knowledge of the breakdown discharge voltage and the breakdown discharge tracks because, particularly in solid materials and for the application of powerful high-voltage sources, the insulation is destroyed in the region of breakdown. Dielectric tests which avoid a breakdown discharge are thus an important aid in the assessment of insulating materials and insulation systems.

1.5.1 Losses in a Dielectric

An ideal dielectric is perfectly loss-free and its behaviour in an electric field can be completely described by a real dielectric constant

$$\varepsilon = \varepsilon_0 \varepsilon_r.$$

Conversely, loss always occurs in a real dielectric. The following may be considered to be the physical reasons for dielectric loss :

Conduction loss P_l by ionic or electronic conduction. The dielectric possesses finite conductivity κ

Polarization loss P_p by orientation, boundary layer, or deformation polarization.

Ionization loss P_i by partial discharges (PD) in or at an insulating arrangement.

These losses cause certain electrical effects, which can be utilized for non-destructive high-voltage tests. The most important measuring quantities are the following:

- conduction current for direct voltage
- dissipation factor for alternating voltage
- partial discharge characteristics for alternating voltage.

For a given test object, these measuring quantities vary with the magnitude of the test voltage. They are also dependent in general upon test conditions, such as temperature and time, as well as upon the properties of the dielectric. Important material parameters are type, composition, structure, purity and previous history. These relationships furnish important information about a dielectric and in special cases may even indicate whether ageing of the dielectric has occurred.

When electrical equivalent circuits are set up for an insulation system, the object is to simulate complicated physical relationships by using circuits with similar electrical properties. However, many dielectric properties are non-linear, and yet one tries to get by with linear circuit elements in the equivalent circuit. These equivalent circuits should therefore be regarded with some caution, and be used only in accordance with their range of validity. Nevertheless, they are of no little use in many problems.

An attempt is made in Fig. 1.65 to present a general equivalent circuit for a dissipative dielectric. Whilst an ideal dielectric may be represented

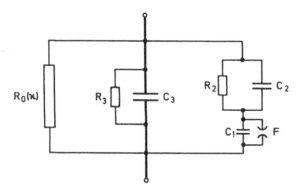

Fig. 1.65 Equivalent circuit of a dielectric with losses due to conduction (κ), polarization and partial discharges

by a pure capacitance C_3, conduction losses can be taken into account by a resistor $R_0(\kappa)$ in parallel. Polarization losses produce a real component of the displacement current, which is simulated by the resistor R_3. Pulse-shaped partial discharges (PD) can be described by the right-hand branch. The gap spacing bridged by a gas discharge is represented by the capacitor C_1 with a gap F in parallel, whereby the repeated recharging of C_1 is effected either by a resistor R_2 or a capacitor C_2. Further details of this PD equivalent circuit will be given in section 1.5.4.

1.5.2   Measurement of the Conduction Current at Direct Voltage

In a steady field of strength $\vec{E}$ the current density $\vec{S}$, according to Ohm's law, for finite specificity conductivity $\kappa = 1/\rho$ is:

$$\vec{S} = \kappa\vec{E}$$

For the specific dielectric loss we then have:

$$P'_{diel} = \vec{E}\,\vec{S} = \kappa\vec{E}^2.$$

The conductivity of insulating systems comprising liquid and solid materials is mostly a result of ionic conduction and is therefore strongly dependent upon temperature, impurities and particularly upon the moisture content. The leakage resistance $R_0(\kappa)$ of an insulating system is determined by measuring the current when a constant direct voltage is applied. Since different types of conduction mechanisms are working simultaneously, the

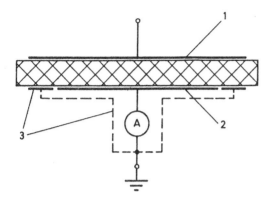

Fig. 1.66 Electrode arrangement to measure the conductivity of an insulation plate at direct voltage
1 High-voltage electrode
2 Measuring electrode
3 Guard-ring electrode and screening

measured result is also time-dependent; and so, for comparable values, the measurement should be performed at a definite time after switching the measuring voltage on, e.g., after 1 minute. For simple electrode geometry the specific quantities κ or ρ can be calculated from the resistance.

A simple arrangement for the investigation of insulating material plates is schematically shown in Fig.1.66. Measuring voltage which is mostly 100 V or 1000 V is applied between electrode 1 and earth. Measuring electrode 2 is earthed through a sensitive ammeter which is directly earthed through the guard-ring electrode 3, absolutely essential for preventing the edge effects and surface currents. In most of the insulating materials κ lies in the region of $10^{-16}...10^{-10}$ S/cm, resulting in measuring currents of the order of picoamperes to nanoamperes. The measuring leads should be appropriately carefully screened.

1.5.3 Measurement of the Dissipation Factor at Alternating Voltage

a) Dielectric Loss and Equivalent Circuits[10]

In an alternating field of strength $\vec{E}$ the current density $\vec{S}$ is generally given by

$$\vec{S} = \left(\kappa + j\omega\underline{\varepsilon} \right)\vec{E}.$$

---

[10] Comprehensive treatment, among others, in *v. Hippel* 1958; *Lesch* 1959; *Anderson* 1964.

As a rule, polarization and ionization losses also occur in a dielectric, besides the conduction loss. The dielectric constant $\varepsilon = \varepsilon_0 \varepsilon_r$ is here no longer a real quantity. The dielectric dissipation factor $\tan \delta$ of an insulation is defined as the quotient of the real component I_w of the current and its imaginary, or reactive, component I_b:

$$\tan \delta = \frac{I_w}{I_b} = \frac{P_{diel}}{P_b}.$$

The dielectric losses comprise three parts, corresponding to the loss mechanism:

$$P_{diel} = P_l + P_p + P_i$$

for each of which an individual dissipation factor can be quoted, so that

$$\tan \delta = \tan \delta_l + \tan \delta_p + \tan \delta_i$$

In a phasor diagram of the fundamental frequency components δ appears as the angle between the current flowing through the dielectric and its imaginary component; for small angles $\tan \delta$ is equal to the power factor $\cos \varphi$.

If only conduction losses occur, then the defining equation gives the simple relationship:

$$\tan \delta_1 = \frac{\kappa}{\omega \varepsilon_0 \varepsilon_r}.$$

This part is inversely proportional to the frequency and can therefore be neglected at high frequencies. For supply frequency, on the other hand, each loss component can have considerable magnitude.

Instead of the dissipation factor, the expression

$$\tan \delta = \frac{\varepsilon''}{\varepsilon'}$$

is often introduced, where ε' is the real part of the relative permittivity and ε'' corresponds to the dielectric loss. Then by definition:

$$P_{diel} = P_b \tan \delta = \omega C \tan \delta \, U^2.$$

If this equation is applied to a volume element, in the form of an infinitesimal cube, for the specific dielectric loss we then have

$$P'_{diel} = \omega \varepsilon_0 \varepsilon' \tan \delta E^2.$$

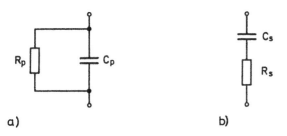

Fig. 1.67 Equivalent circuits for a dissipative dielectric for alternating voltage
a) parallel equivalent circuit, b) series equivalent circuit

When this equation is applied, the approximation $\varepsilon' \approx \varepsilon_r$ can generally be used.

Fig. 1.67 shows two equivalent circuits for a dissipative dielectric at alternating voltage. Here the losses are simulated by resistances. The dissipation factor is given for the parallel circuit of Fig. 1.67a by :

$$\tan \delta = \frac{1}{R_p \omega C_p}$$

and for the series circuit of Fig. 1.67b by :

$$\tan \delta = R_s \omega C_s.$$

For a fixed frequency, both equivalent circuits are indeed equivalent and the elements can be converted accordingly. However, the frequency dependence is just the opposite in the two cases, and this illustrates the limited validity of these equivalent circuits.

With certain restrictions, even the frequency dependence can be simulated by adding other elements to these circuits.

b) Measurement with the Schering Bridge
Measurement of the dielectric loss at alternating voltage is usually carried out in high-voltage technology using the bridge circuit devised by *H. Schering* in 1919 (Fig. 1.68).The Schering bridge is an alternating current bridge made up of capacitors and resistors. The balancing elements to be adjusted are inside an earthed enclosure whilst the test object C_x and the standard capacitor C_2, which should be practically loss-free, are at high voltage. The null indicator N may be sensitive only to the fundamental frequency of the test voltage, which usually deviates from the sinusoidal.

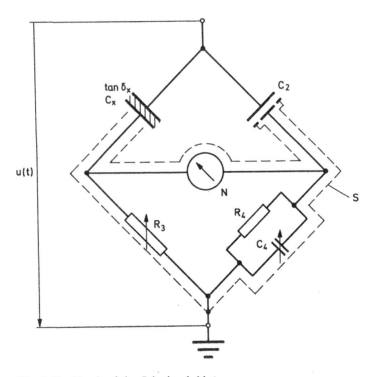

Fig. 1.68 Circuit of the Schering bridge
C_x Test object, C_2 Standard capacitor, R_3, C_4 Balancing elements,
R_4 Fixed resistor, N Null indicator, S Screening

The corner points of the bridge must be provided with overvoltage protective devices so that overvoltages in the low-voltage circuit will be prevented, should the test object breakdown.

The capacitance and dissipation factor of the test object are determined by adjusting the resistor R_3 and the capacitor C_4. For the balanced bridge, the following relation holds for the admittances of the bridge arms:

$$\underline{Y}_4\underline{Y}_x = \underline{Y}_2\underline{Y}_3$$

Taking the parallel equivalent circuit of Fig. 1.67a for the test object and substituting we have:

$$\left(\frac{1}{R_4} + j\omega C_4\right)\left(\frac{1}{R_x} + j\omega C_x\right) = \frac{j\omega C_2}{R_3}.$$

This equation must be satisfied by the real as well the imaginary components, which corresponds to the Schering bridge balanced in amplitude and phase. The following relationships result :

$$\frac{1}{\omega C_x R_x} = \omega C_4 R_4$$

$$\frac{C_4}{R_x} + \frac{C_x}{R_4} = \frac{C_2}{R_3}$$

For the required quantities we have :

$$\tan\delta_x = \omega \, C_4 R_4$$

$$C_x = C_2 \frac{R_4 \, / \, R_3}{1 + \tan^2 \delta_x} \approx C_2 \frac{R_4}{R_3}$$

Reflecting the significance of dissipation factor measurements at high voltages, the circuits and the equipment required to carry out these measurements have been the object of considerable further development compared with the basic circuit described above. One may mention here the additional circuitry to compensate the earth capacitances at bridge corners [*Potthoff, Widmann* 1965; *Schwab* 1969].

The null indicator should record currents corresponding to the fundamental frequency of the alternating test voltage only, since otherwise correct balancing is impossible; more sensitive electronic null indicators with oscillographic indication are preferred instead of mechanical vibration galvanometers.

The realization of the standard capacitor C_2 is a high-voltage technological speciality. For the derivation of the balanced condition relation, it was assumed, *a priori* , that the dissipation factor of the standard capacitor is negligibly small compared with that of the test object. One therefore chooses capacitors with gas as an especially low loss dielectric. An arrangement with coaxial cylinder electrodes and compressed gas insulation, suggested in 1928 by *H. Schering* and *R. Vieweg*, is found to be particularly suitable for high voltages. Exact knowledge of the capacitance C_2 of the standard capacitor is essential for measurement of C_x. C_2 should therefore be unaffected by external influences as far as possible [*Latzel, Schon* 1987].

c) Transformer based Current Comparison Methods

Instead of the Schering bridge even a transformer based current comparison circuit is also utilised. Fig.1.69 shows such an executed circuit [*Kuffel, Zaengl* 1984]. The differential transformer with the windings L_1, L_2 and L_N

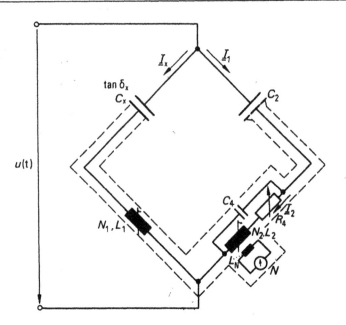

Fig. 1.69 Transformer based current comparison circuit
C_x, C_2, N, S see Fig. 1.68, N_1, N_2, R_4 balancing elements, C_4 fixed capacitor

permits a finely stepped adjustment of the number of turns N_1 and N_2. The differential flux is balanced with the null indicator. The balance condition for a close magnetic coupling of the windings reads:

$$I_x \cdot N_1 = I_2 \cdot N_2$$

Substituting, one obtains making use of the series equivalent circuit of Fig.1.67b for the specimen:

$$\frac{N_1}{R_x + \dfrac{1}{j\omega C_x}} = \frac{N_2 \cdot j\omega C_2}{1 + j\omega R_4 (C_4 + C_2)}$$

For the required parameters we have:

$$\tan \delta_x = \omega R_4 (C_4 + C_2)$$

$$C_x = C_2 \frac{N_2}{N_1}$$

The advantage of the transformer based bridge lies in the smaller effect of the stray capacitances of the measuring cable.

1.5.4   Measurement of Partial Discharges at Alternating Voltages[11]

When the breakdown discharge strength of an insulation system is locally exceeded, either full or partial breakdown occurs, depending upon whether a discharge channel of low resistance develops between the electrodes or not. In the case of a partial breakdown the conditions for a stationary discharge are often not fulfilled as a consequence of insufficient energy input, and so only a short discharge pulse results. If the dielectric possesses self-healing properties in the overstressed region, and if the electric field builds up again, pulse-shaped partial discharges occur. Partial discharges may however also be pulse-free when they occur in conjunction with a stationary gas discharge.

Pulse-shaped and pulse-less partial discharges can occur with any type of voltage. However, with regard to the practical significance, only partial discharges at alternating voltages shall be discussed below, although the treatment of external partial discharges in particular is also valid for direct voltages.

a) External Partial Discharges

Collision ionization sets-in in gases when the onset voltage is exceeded at electrodes of small radius of curvature. In a strongly inhomogeneous field electron avalanches and photo-ionization produce imperfect breakdown channels which, for alternating voltage, must re-ignite after each extinction of the partial discharges at the voltage zero-crossover. This phenomenon, denoted external partial discharge or corona discharge, has great practical significance, above all for overhead high-voltage transmission lines, because the maintenance of the discharge uses up energy (corona losses) and the current pulses produced generate electromagnetic waves (radio interference). External partial discharges also occur in test circuits for non-destructive high-voltage testing, and impede the detection of internal partial discharges which could possibly endanger the dielectric. External partial discharges can also occur in liquid insulators or at the interfaces of solid insulating materials; in the long run they will cause weakening of the insulation there, which might subsequently lead to a complete breakdown.

Fig. 1.70 shows a point-plane electrode configuration as a typical example of a setup with external partial discharges, as well as a grossly simplified equivalent circuit for pulse-shaped partial discharges. C_1 represents the

---

[11] Summarising treatment in *Kreuger 1964; Schwab 1969; Stamm, Porzel 1969; Bartnikas, MeMahon 1969; Kuffel, Zaengl 1984; Beyer et al. 1986.*

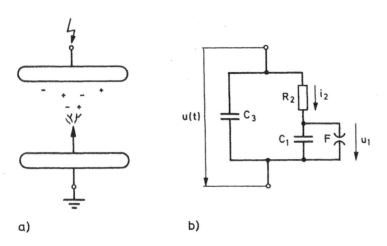

Fig.1.70 Arrangement with external partial discharge and equivalent circuit
a) point-plane electrode configuration, b) equivalent circuit

capacitance assigned to the respective gas space breaking down and will be completely discharged when the ignition voltage U_z of the gap F is reached. The charge carriers formed at the point wander into the field, causing a certain conductivity which is represented in the equivalent circuit by R_2. Finally, C_3 is a parallel capacitance given by the electrode arrangement. Assuming that $R_2 \gg 1/\omega C_1$, the current through R_2 is:

$$i_2 = \frac{u(t)}{R_2} \cdot$$

If, as drawn in Fig. 1.71,

$$u(t) = \hat{U} \sin \omega t$$

is written for the test voltage, the open circuit voltage at C_1 at the end of the transient phase would be:

$$u_{10} = \frac{\hat{U}}{\omega C_1 R_2} \sin(\omega t - \pi / 2).$$

If the peak value of the test voltage just attains the onset voltage

$$\hat{U}_e = \omega C_1 R_2 U_z,$$

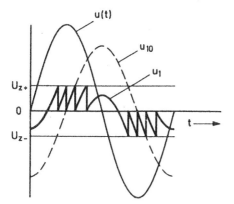

Fig. 1.71 Voltage curves in the equivalent circuit for pulse-shaped external partial discharges

then the ignition voltage U_z appears across F and C_1 will be discharged in one burst. For increasing test voltage $u(t)$, C_1 will subsequently be recharged by a voltage parallel in shape to u_{10} until U_z is reached once more, and so on. From the curves drawn for the voltage u_1, one can see that the partial discharge impulses occur predominantly in the peak of the test voltage. The pulse repetition rate n as a function of $\hat{U}$ is shown in Fig. 1.72. The dashed straight line represents a good approximation for high pulse repetition rate:

$$n \approx 4f\frac{\hat{U} - \hat{U}_e}{\hat{U}_e}.$$

During each individual discharge, a quantity of charge

$$Q_1 = C_1 U_z = \frac{\hat{U}_e}{\omega R_2}$$

is compensated for in F. This charge is then fed to C_1 again via R_2 by the voltage source and can be recorded as:

$$\Delta Q = Q_1.$$

The prefix Δ is intended to show that this is a quantity corresponding to a complete pulse, yet with duration very short by comparison with the test voltage period.

The given equivalent circuit, by adding extra elements, can be made to correspond better to the physical behaviour of a real setup with external partial discharges. For example, one should note that the value of the onset

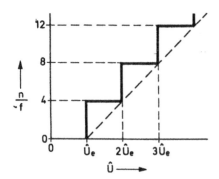

Fig.1.72 Partial discharge pulse repetition rate

voltage depends in many cases upon the polarity. The addition of a rectifier parallel to C_1 enables periodic discharges of only one polarity to be considered. The rectifier provides for the discharge of the capacitor C_1 during the half-cycles in which no partial discharges occur, because the onset voltage is too high.

b) Internal Partial Discharges

Where cavities are present in the liquid or solid dielectric of insulation systems, the field strength is greater inside than in the surrounding medium. When the voltage across the cavity exceeds the ignition voltage, a partial breakdown results. Particularly for alternating voltages of sufficient amplitude a normally pulse-shaped discharge occurs in the cavity. The surrounding dielectric may deteriorate due to the long-term effect of these internal partial discharges, and under certain circumstances it can even be destroyed by total breakdown as a result of an erosion mechanism.

As a typical example of an electrode arrangement with internal partial discharges Fig. 1.73 shows an insulation system with a solid dielectric containing a gaseous cavity. The figure also shows the equivalent circuit for pulse-shaped partial discharges, suggested in 1932 by *A. Gemant* and *W.v.Phillippoff*. C_1 corresponds to the cavity capacitance which discharges completely via F when the ignition voltage U_z is reached. C_2 corresponds to the capacitance in series with the cavity and C_3 is the parallel capacitance of the arrangement.

For a sinusoidal test voltage the open circuit voltage across C_1 is given by

$$u_{10} = \frac{C_2}{C_1 + C_2} u(t) = \frac{C_2}{C_1 + C_2} \hat{U} \sin \omega t$$

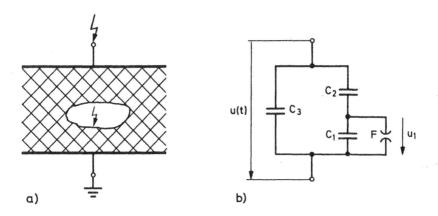

Fig. 1.73 Arrangement with internal partial discharge and equivalent circuit
a) test object with cavity, b) equivalent circuit

The peak value of the test voltage reaches the onset voltage $\hat{U}_e$ when the peak value of the open circuit voltage is just equal to U_z. It then follows that

$$\hat{U}_e = \frac{C_1 + C_2}{C_2} U_z .$$

If the test voltage is greater than the onset voltage, repeated charging of C_1 occurs, as shown in Fig. 1.74. It may be recognised that the partial discharge pulses occur predominantly in the region of the test voltage zero-crossover. Here too, the same relationship holds for pulse repletion rate as already given in a) and shown graphically in Fig. 1.72. The different phase relation of external and internal partial discharges is an important distinguishing characteristic of these two phenomena.

The charge compensated at the discharge site for each discharge is:

$$Q_1 = (C_1 + C_2)U_z,$$

while only the "apparent" charge Q_{1s} is fed to C_2, where

$$Q_{1s} = C_2 U_z \neq Q_1 .$$

As opposed to an arrangement with external partial discharges, it is therefore basically impossible to measure the real charge Q_1 with unknown partial capacitances. The charge Q_{1s} which can be measured is, analogous to 1.5.4a, designated as ΔQ:

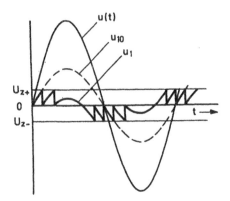

Fig. 1.74 Voltage curves in the equivalent circuit for pulse-shaped internal partial discharges.

$$\Delta Q = Q_{1s} = \frac{C_2}{C_1 + C_2} Q_1 \cdot$$

The only measurable apparent charge ΔQ is proportional to the transferred charge Q_1 at the void. The proportionality factor can however be determined only for known capacitances C_1 and C_2. The size and the shape of the void are generally unknown so that a quantitative measurement is not possible in practice. This makes the evaluation of partial discharge parameters measured so difficult.

c) Measurement and Recording of Partial Discharge
In Fig.1.75 a PD test circuit is shown which is an equivalent circuit valid for external and also internal partial discharges. A generator supplies in the short time $\Delta t \to 0$, a current pulse with the impressed charge ΔQ; the pulse acts initially only on the test object capacitance C because of the impedances in the connecting leads. With the source current i_Q, we have:

$$\Delta Q = \int_{\Delta t \to 0} i_Q dt = const.$$

Due to this the voltage across the test object suddenly changes by an amount

$$\Delta U = \frac{\Delta Q}{C} \cdot$$

This represents a very rapid process inside the test object, since the presented model contains no elements which could delay recharging.

For the subsequent transient processes in the test circuit only the capacitor C_k of the high-voltage circuit and the measuring resistor R are of significance. Referring to Fig. 1.75, one may now consider which of the partial discharge quantities can be measured at R. For the transient processes the nodal equation is:

$$i_{TE} = i_C + i_Q$$

and the loop equation is:

$$i_{TE}R + \frac{1}{C_k}\int_0^t i_{TE}dt + \int_0^t i_C dt = 0.$$

By eliminating i_C it follows for the current pulse with impressed charge supplied by the generator:

$$\Delta Q = \left(1 + \frac{C}{C_k}\right)\int_0^t i_{TE}dt + i_{TE}RC.$$

The condition ΔQ = constant must be satisfied at all times t; this requirement is fulfilled by the Dirac pulse assumed for i_Q. For $t \to 0$ it follows from the above equation that i_{TE} jumps to a final value at the instant of the pulse. Since there is no contribution from the integral at this instant, the initial value of i_{TE} is obtained from:

$$\Delta Q = i_{TE}RC = \Delta U_{TE}C.$$

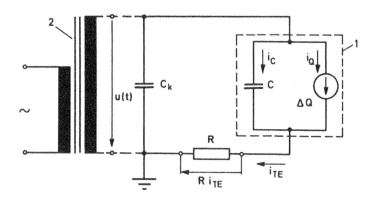

Fig.1.75 Test circuit for measuring partial discharge
1 Specimen with external or internal PD, 2 Test transformer, R Measuring resistor , C_k Coupling capacitor

A voltage jump ΔU_{TE} thus appears at the measuring resistor R. For $t \rightarrow \infty$, i.e. for times after decay of the transient process, the product $i_{TE}RC$ vanishes and we have:

$$\Delta Q = \left(1 + \frac{C}{C_k}\right) \int_0^{t \to \infty} i_{TE} \, dt = \left(1 + \frac{C}{C_k}\right) \Delta Q_{TE}.$$

Here ΔQ_{TE} represents the pulse-shaped charge flowing through R. Most of the measuring methods evaluate quantities which can be determined at the measuring resistor R when pulse-shaped partial discharges appear. When C is known, ΔQ may be determined from ΔQ_{TE}; with the additional knowledge of C_k, ΔQ may also be determined from ΔQ_{TE}. The expression $(1 + C/C_k)$ corresponds to a transmission ratio of significance to the measuring sensitivity.

When partial discharge measurements are performed, a special coupling capacitor may be taken for C_k and the measuring resistor can then be connected in the earth lead of C_k. However, the high-voltage side capacitances of the system, especially the winding capacitances of the transformer, are usually of sufficient magnitude ,so that one can do without a special coupling capacitor. For an adequate value of the coupling capacitance $(C_k \gg C)$ the measured charge ΔQ_{TE} becomes equal to the charge ΔQ which is fed to that part of the insulation containing the partial discharge site. The transmission ratio may be experimentally determined in the setup with the aid of a pulse generator supplying impressed charge pulses [*Mole* 1970, *Schon* 1986].

Since with ΔU_{TE} and ΔQ_{TE} we are in fact dealing with statistically varying processes, it is usual to determine the mean value of the measured quantities in these measurements (VDE 0434; IEC Publ. 270). Under favourable conditions, knowing the trends of certain partial discharge parameters such as charge, pulse repetition rate, energy and time-dependence of the partial discharge pulses during stressing, the effect of internal ionisation processes on the quality of an insulating system can be assessed. On the other hand, one may not expect to make a reliable prediction of the life expectancy of an insulating system based upon a single measurement only [*Leu* 1966; *Kind, König* 1967; *Kodoll* 1974].

In testing practice, measurement of the apparent charge of the PD pulses as the measured parameter during PD tests is invoked. In spite of the influencing of PD pulses by the test circuit and the measuring system, the apparent charge is correctly determined together with a calibration [*Schon* 1986].

For measurement of the apparent charge of PD pulses, equipment based on various working principles are available. Under the assumption that the PD pulse of the apparent charge, as a rule, has a constant amplitude density at the specimen terminals up to about 500 kHz. narrowband ($\Delta f = 10$ kHz) and wideband (50 kHz ...500 kHz) PD measuring instruments which carry out the integration in frequency domain can be made use of. Narrowband equipment have the advantage that one can by-pass disturbances in a definite frequency domain; the disadvantage lies in the fact that PD pulses can no longer be distinguished as to their polarity. Wideband measuring equipment supply in contrast, the polarity information also. If the PD current pulse is measured as a very wideband (> MHz) one, an integration in the time domain can also be carried out.

Besides the electrical PD measuring technology on the basis of integration of PD current pulses, for special applications, methods are applied which evaluate the acoustic or optical wave propagation from the void [*Feser* 1992]. Measurement of the dielectric dissipation factor yields evidence about the losses occurring in the total electrically stressed volume; here, apart from the basic losses P_l+P_p, the partial discharge losses P_i too are determined. The considerations above using the equivalent circuit apply to pure pulse-shaped discharges. With the dissipation factor measurement, unlike many other measurements, one also obtains the loss contributions caused by partial discharges which are not pulse-shaped. However, the dissipation factor measurement is usually not sufficiently sensitive to locate single partial discharge sites in an insulating system.

2 Layout and Operation of High-Voltage Test Setups

2.1 Dimensions and Technical Equipment of the Test Setups

The dimensions and equipment of a high-voltage laboratory[1] are primarily determined by the magnitude of the voltage to be generated. A second important feature is the intended application e.g., for teaching purposes, as a testing or research laboratory.

2.1.1 Stands for High-Voltage Practicals

Practicals are laboratory exercises which give the students an opportunity to conduct set experiments under supervision. The experiments would generally be performed in small groups of three up to a maximum of six participants. The experimental stands described below are designed for this kind of practicals.

In order to accommodate a large number of students, more stands must be available in which experiments can be conducted simultaneously. A useful guide for the setup of a practical laboratory would be approximately 1 experimental stand for every 20 students. The number of stands so derived imposes a certain restriction upon the voltage amplitude for economic reasons, which is also expedient with regard to clearer arrangement, and with that safety, of smaller setups.

If the maximum alternating voltage is restricted to 100 kV and the power ratings to between 5 and 10 kVA, the experimental stands could be set up

---

[1] Comprehensive treatment, among others, in *Marx* 1952; *Prinz* 1965; *Craggs, Meek* 1954; *Leroy, Gallet* 1975; *Hylten-Cavallius* 1986.

in rooms with a normal height of 2.5 m. Moreover, the weight of the required construction elements, with the exception of the testing transformer, would be low enough to allow transport without crane facilities. Since most of the basic physical phenomena can already be observed within a voltage range of about 100 kV a.c., the restriction on this value does not impose any appreciable limit on the choice of experiments to be carried out. If necessary, the scope of the practicals could be widened by some demonstration experiments at a very high voltage.

As a proven example, one of the five identically set up experimental stands for high-voltage practicals at the High-Voltage Institute of the Technical University, Braunschweig will be described on the basis of Fig. 2.1. The protective barrier 1, consisting of wire mesh fixed to a metallic frame, is provided with a lockable door, near which work table 3 and control desk 4 are arranged. Inside the barrier is a working platform of two welded steel frames 5 with a covering of four hardwood panels. The steel frames serve as earthing points.

The high-voltage circuits are set up on the working platform; construction elements and accessories which are not required may be stored in trolley-drawers underneath. Further details may be taken from Fig. 2.2. For example, flexible cables for control and measuring purposes are already laid between the control desk and the working platform and need only be connected to the construction elements.

A commonly adopted construction dispenses with the working platform. The high-voltage circuits would be set up directly on the ground, which is provided with an earthing sheet of aluminium or copper. Heavier circuit elements (e.g. transformers) or some parts of circuits could be mounted on trolleys.

2.1.2   High-Voltage Testing Bays

In planning these, it should be considered that they are often responsible for an appreciable portion of the total investment and personnel costs. The introduction of partially automated measuring and protocolling devices permits considerable saving of costs. Testing bays where routine and type tests on manufactured high-voltage equipment are to be carried out, are usually adapted to a particular kind of test. They should constitute an integral part of the production line.

In practice, the operating voltage of the equipment to be tested influences the spatial arrangement of the testing bay. This is because, with respect to

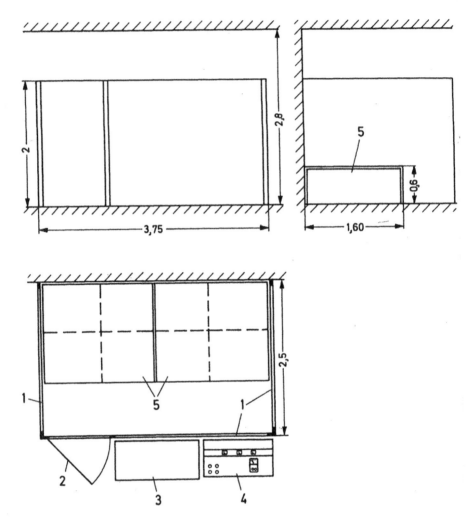

Fig. 2.1 Dimensions of experimental stands for high-voltage practicals (in m)
1 Protective barrier, 2 Door, 3 Working table
4 Control desk, 5 Working platform

the minimum clearances to be maintained in a test setup, the clearance of the test room must, for high voltages, be considerably greater than the height of the production rooms. In consequence, a test object with an operating voltage over 245 kV would, for constructional reasons alone, require its own testing hall. In this case, electromagnetic screening and

Fig. 2.2 Experimental stand for a high-voltage practical
(Dimensions according to Fig. 2.1, Photo: E. Sitte, Braunschweig)

darkening facilities for the room could also be easily introduced. Fig. 2.3 shows, as an example the possible layout of the testing hall for a factory in which transformers up to 400 kV are produced.

Numerous technical disadvantages prevail when carrying out high-voltage tests in open air. In addition, there is uncertainty in planning the test schedule due to weather conditions, which is generally intolerable, particularly in factory testing bays. As a rule therefore, one prefers an indoor solution.

The high-voltage installations for testing bays conform to the maximum test voltage required, as well as the load represented by the test object. The guiding values listed in Table 2.1 are intended to give an idea of the magnitude of the test voltages for 3-phase high-voltage equipment with a given operating voltage. In any specific case, the exact values have to be taken from the appropriate test specifications valid at the time.

When selecting the voltage generators using Table 2.1, one should observe that the rated voltage of the generator must be chosen to be higher than the test voltage given in the table. For a.c. testing transformers, an increase of about 10 % of the required test voltage is sufficient. For impulse voltage

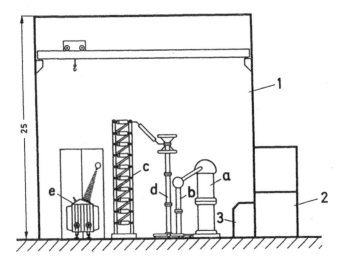

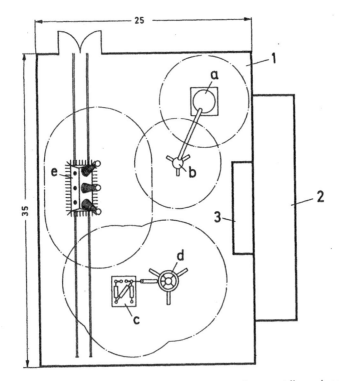

Fig. 2.3 Testing bay for 400 kV power transformers (dimensions in m)
1 Screened high-voltage hall : a testing transformer (800 kV).
　b capacitor (800 kV). c impulse generator (3 MV).
　d impulse voltage divider (3 MV). e test object
2 Adjacent testing rooms
3 Control desk

generators it is usual to specify the total charging voltage, which must be multiplied by the utilization factor to derive the peak value of the impulse voltage. The utilization factor is, however, influenced by the test object and, above all for the generation of switching impulse voltages, it can even assume values below 0.5. The following guiding values may be given for the factor by which the highest required withstand voltage should be multiplied in order to derive the required rated voltage of the test voltage generator:

Alternating voltage	1.1
Lightning impulse voltage	1.8
Switching impulse voltage	2.0
Direct voltage	1.2

Table 2.1   Guiding values for test voltages and minimum clearances for test setups

Operating voltage kV	Alternating voltage kV	Lighting Impulse voltage kV	Switching Impulse voltage kV	Minimum clearance m
36	70	170	-	
123	230	550	-	
245	460	1050	-	
420	510	1425	1050	5
525	630	1550	1175	8
765	850	2100	1550	12
1200	1400	2550	2100	20
1600	1900	3150	2550	30

The minimum clearances listed in the last column of Table 2.1 are guiding values for the lowest required air spacings between the parts of the test setup at high-voltage potential and those points in the surroundings at earth potential. At very high operating voltages the minimum clearances are determined by the magnitude of the positive switching impulse voltage to be generated, since a given electrode configuration can have a particularly low breakdown voltage for this type of voltage.

The majority of high-voltage test objects represent capacitive loading of the test voltage source. The following guiding values may be given for the capacitances [*Siemens* 1960]:

Supports, insulators		20 pF

Bushings, inductive instrument transformers		200-400 pF
Power transformers (high-voltage winding against all other parts)	up to 1 MVA	3000 pF
	up to 100 MVA	25000 pF

Cable sample upto 10 m long		3000 pF

Experimental setup, measuring capacitor, Leads for a.c. test voltage	up to 100 kV	100 pF
	up to 1000 kV	1000 pF

The more common features of testing transformers are compiled in Table 2.2. Testing transformers are designed in most cases for short-time service of 15 min to 1 h; continuous operation is necessary only for temperature-rise tests or for an investigation of dielectric stability (cables, bushings).

Table 2.2 Examples for the design data of testing transformers

Rated voltage	Rated current	Rated power	Short-circuit impedance
kV	A	kVA	%
100	0.1	10	10
300	0.3	100	10
800	0.5	400	15
1200	1	1200	25
2000	2	4000	25

The corresponding data for impulse voltage generators are contained in Table 2.3. Usual values of the impulse capacitance lie at a few tens of nF. The table was compiled using an average value of $C_s = 25$ nF for the calculations. For test objects which represent particularly heavy loading (cables, power transformers), it may become necessary to choose an even larger impulse capacitance [Widmann 1962].

Observance of the safety regulations must be enforced with particular rigour in testing bays, since the routine nature of the work very often results in a slackening of care, even in qualified personnel. Moreover, unqualified personnel also have to enter the danger-zone, particularly during delivery and return of the test objects. Organizationally well thought out and technically reliable safety measures must prevent any risk due to the high voltage.

Table 2.3 Examples for the design data of impulse voltage generators

Total charging voltage U_0 kV	Impulse capacitance C_s nF	Impulse energy W kWs	W/U_0 kWs/MV
200	25	0.5	2.5
400	25	2.0	5
1000	25	12.5	12.5
2000	25	50	25
4000	25	200	50

Further details of established high-voltage testing bays can be found in the literature [*Elsner* 1952; *Gsodam, Stockreiter* 1965; *Läpple* 1966; *Hevne* 1969; *Raupach* 1969; *Hylten-Cavallius* 1986].

2.1.3   High-Voltage Research Laboratories

Some reference data for the selection of basic equipment for high-voltage laboratories intended for research and development work are contained in the information given for testing bays in Tables 2.1 to 2.3. It is clear from the nature of the assignment that no further details can be given about the data of technical equipment required for specific research projects. The following exposition shall be restricted to mentioning the overall aspects and drawing attention to the description of established practical setups in the literature.

Apart form meeting the technical requirements, the planning of research laboratories must also provide for the greatest possible flexibility. In the arrangement of test rooms with different dimensions and facilities, any non-essential commitment to a particular application should be avoided. Besides this, it is desirable that even the largest rooms, which are designed for the highest voltages, should permit temporary subdivision in order to ensure simultaneous use. By experience the highest voltage is only rarely required.

Moreover, for high-voltage laboratories in particular, it can also be useful to provide for a multipurpose experimental hall with an average height of 6...12m and floor area of a few hundred m^2; this hall can then be subdivided into a large number of experimental stands using movable metallic barriers, depending upon the respective projects. If appropriate precautions are taken, the advantages of this kind of flexibility outweigh the disadvantages of mutual disturbances.

It is convenient if the experimental hall with the highest voltage borders on an outdoor site where unwieldy test objects in particular can be investigated, using the voltage source of the hall. Wall bushings for test voltages above 500 kV a.c. can only be produced at great expense. In addition difficulties arise with regard to safety and measuring techniques, on account of the optical separation between the voltage source and the test object, as well as due to the self-capacitance of the bushing. It would in general be better, therefore, to provide the experimental hall with a large gateway through which either the voltage sources may be temporarily taken out into the open, or through which the test voltage is led out by means of a simple wire connection.

When planning high-voltage laboratories for research and development work one should make certain that a sufficient number of auxiliary rooms are provided. Apart from the obvious rooms for offices, workshops, power supply, etc. one should above all not forget rooms for storage, stowing and packing. With a height of 3 to 6 m, these rooms should together take up at least one third of the floor area of the laboratory itself. If one economizes on this point, it is likely that a large portion of valuable laboratory space will be blocked up by temporarily unused testing and auxiliary equipment after a short working period. Of course, good transport facilities must exist between the testing rooms and stowing areas. Coverable loading hatches in testing rooms within reach of the crane equipment and which provide access to the stowing rooms located in the cellar, have been found very useful.

Further particulars of established high-voltage laboratories can be found in the literature [*AEG* 1953; *Micafil* 1963; *Prinz* 1965; *Leschanz, Oberdorfer* 1968; *Nasser, Heiszler* 1965; *Leroy et al.* 1971].

2.1.4   Auxiliary Facilities for Large Test Setups

Under auxiliary facilities, we understand here crane equipment, means of transport, heating, illumination etc. For testing bays in the production lines of factory halls, the choice of most these auxiliary facilities would already be made by the mode of production.

The selection of crane equipment is important; this can take the form of either bridge cranes, overhead tackles or fixed lifting gear. For loads over 50 t, only bridge cranes can be considered from the constructional point of view. These can indeed reach any part of the hall, but introduce the inherent disadvantage that suspension of elements of the experimental setup from the ceiling impedes the movement of the crane bridge. This suspension is particularly convenient for voltage carrying parts, since otherwise one would be forced to use support insulators, insulated stands or insulating wall-spacers to set up the experimental circuit. This disadvantage does not arise for the proven arrangement of several 'overhead tackles for medium loads (2...10 t). These allow experimental elements to be fastened to the ceiling by means of hooks or fixed pulleys, and can themselves be used to suspend certain components. An insulated suspension for light components, such as the high-voltage connections, can be effected with plastic ropes; for heavier loads glass-fibre reinforced plastic rods or suspension insulators are suitable.

An important decision to be taken in view of future test objects concerns the necessary loading capacity of the research room floors. This lies in the range of $0.5...2$ t/m^2, and is essentially influenced by the question whether the large-scale equipment is mobile or not. Mobility should definitely be aimed at from the standpoint of optimal utilisation of the rooms. Loads over about 10 t can be placed or moved on rails embedded in the floor. Since the freedom of installation is restricted by permanent rails, preference should normally be given to the rather more expensive solution without fixed rails. Here for heavy loads air-cushioned foundations have proved their merit. Loads up to a few 10t can also be made mobile at minimum expense by means of rails laid openly on the floor and interchangeable flat rollers on the apparatus base.

Regarding the heating of the testing rooms, the requirement is that it should function dust-free and noiselessly. The illumination in the testing room, which should if possible be able to be completely blacked out, must be finely adjustable, since visual observations and optical measurements represent an indispensable auxiliary aid to high-voltage experiments.

Further details of auxiliary facilities for high-voltage research laboratories can be found in the literature [*Marx* 1952; *Prinz* 1965; *Leroy, Gallet* 1975].

2.2 Fencing, Earthing and Shielding of Test Setups

The equipment for fencing, earthing and shielding research laboratories for high voltage is intended to prevent risk to persons, installations and apparatus. At the same time undisturbed measurement of rapidly varying phenomena should be ensured and undesirable mutual interference between experimental setup and the environment avoided.

2.2.1 Fencing

The actual danger-zone of the high-voltage circuit must be protected against unintentional entry by walls or metallic fences. Simple barrier-chains or the identification of the danger zone solely by warning signs can be considered sufficient only where their observance can be constantly supervised. Entrances to the danger-zone should be provided with locks which effect automatic switch-off.

Visible metallic connection with earth must be established before the high-voltage elements are touched. In case of smaller experimental setups, such as a practical, this can be done before entering the setup with the help of insulated rods introduced through the fencing mesh, and which establish the ground connection inside. In larger setups placing the earthing rod should be the first action after entering the danger zone, or automatic earthing switches should be provided. Complete earthing is especially important when the circuit contains capacitors charged by direct voltage. Further particulars concerning this topic may be found in Appendix 4.1 under Safety Regulations.

2.2.2 Earthing Equipment

Apart from the obvious measures to guarantee reliable earth connection for steady working conditions, one must remember that rapid voltage and current variations can occur during high-voltage experiments as a result of breakdown processes. In consequence, transient currents appear in the earth connections and these can cause potential differences of the same order of magnitude as the applied test voltages.

Elements at earth potential during steady operation can temporarily acquire a high potential, though in general personal risk is not the consequence. On the other hand, damage to equipment and disturbance of measurements often takes place. The reasons for these phenomena and measures to suppress them shall be discussed briefly in the following [*Stephanides* 1959; *Möller* 1965; *Sirait* 1967; *Hylten-Cavallius, Giao* 1969; *Lührmann* 1973].

The sudden voltage collapse on breakdown discharge occurs in such short times, that even lightning impulse voltage appears to be slow by comparison. The discharge which develops at the breakdown discharge site is, to a first approximation, fed by discharge of a capacitor, which, in the case of an impulse voltage generator is essentially the load capacitance and, for a testing transformer, the capacitances of the high-voltage winding and the test setup.

Significant properties of this state of affairs can be inferred from the simple scheme of Fig. 2.4. A capacitor C_a begins to discharge at $t = 0$ through a gap which can be bridged in a very short time. The electrical behaviour of the circuit can be described by means of the equivalent circuit shown, where L_a denotes the inductance of the entire chopping circuit. The result is the indicated periodically damped form of the capacitor voltage u_c and current i_a. The first immediately recognizable requirement is that the path of the current i_a should be closed in an appropriate circuit.

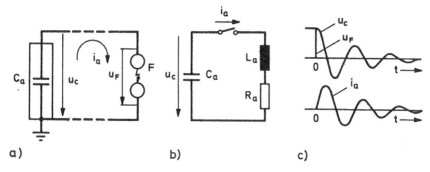

a)   b)   c)

Fig. 2.4 Simple scheme of a high-voltage circuit with breakdown discharge process a) circuit setup, b) equivalent circuit, c) voltage and current curves

An electric field develops between the elements at high-voltage potential and the neighbouring elements at earth potential. This stray earth field can be simulated by a distributed earth capacitance C_e, as shown in Fig. 2.5a [*Sirait* 1967]. Should C_a be rapidly discharged, a transient earth current i_e is produced, generated by the potential variation of the chopping circuit; this current flows at least partly outside the test setup and may cause undesirable overvoltages there. If on the other hand the entire high-voltage circuit is surrounded by a closed metallic shield, a Faraday cage as in Fig. 2.5b, then the earth currents too flow in predetermined paths and the earth connections outside the cage remain current-free. These earth connections can therefore be designed exclusively according to the requirements of adequate steady-state operational grounding.

As a rule, the floor of the laboratory, as shown in Fig. 2.6, is covered, at least within the region of the high-voltage apparatus, by a plane earth conductor with as high a conductivity as possible (foil or closely meshed copper grid); the earth terminals of the apparatus are connected to it non-inductively using wide copper bands, to keep the voltage drops due to large currents at a minimum. All measuring and control cables, as well as

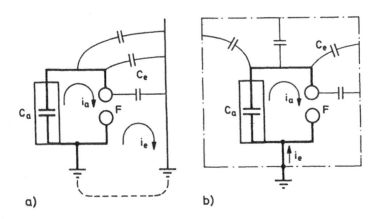

Fig. 2.5 Earth currents in high-voltage setups
a) without shielding. b) with a Faraday cage

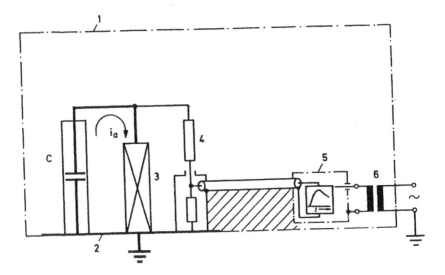

Fig. 2.6 Earthing and shielding of a high-voltage test setup
1 Faraday cage
2 Reinforced plane earth conductor on the floor
3 Test object
4 Voltage divider
5 TR enclosure or measuring cabin
6 Power supply to TR via isolating transformer, and low-pass filter when necessary

earth connections, should be laid avoiding large loops and if possible even run underneath the plane earth conductor, e.g., in a metallic cable duct [*Kaden* 1959; *Schwab* 1969].

Particular attention should be paid to the connection of the oscilloscope if disturbances during the measurement of rapidly varying phenomena are to be avoided. The measuring signal is always transmitted to the measuring device via shielded cables – coaxial measuring cables as a rule. Hereby one should however prevent currents which do not return in the inner conductor, from flowing in the earthed sheath of the measuring cable, since the corresponding voltage drop is superimposed on the measuring signal as an interference voltage.

Interfering sheath currents can be prevented in various ways. If possible the highly conductive plane earth conductor on the laboratory floor is made large enough so that the cable sheath, adjacent to it and connected with it electrically at several points, is relieved of disturbing currents. Better still is the installation of the measuring cable outside the Faraday cage, e.g. in a metallic cable duct or in an earthed metal tube. A replacement of the measuring cable with fibre-glass connection and optical transmission of the measured signal would be interference-proof.

It is however often unavoidable that the measuring cable sheath and the earthing system form a closed loop in which disturbing circulating currents may flow as a result of rapidly varying magnetic fields. An earth loop of this kind is indicated by the hatching in Fig. 2.6 for the example of a voltage measurement. The transient recorder enclosure in this case may only be connected by the measuring cable to the earth end of the divider and earthed; that is also why the line input to the transient recorder has to be fed via an isolating transformer. This isolating transformer is expedient in any case since potential differences are always present between the earth lead of the mains and the earthed parts of the experimental setup. The disturbing sheath currents may be reduced further by placing ferrite cores over the measuring cables.

In high-voltage setups with very rapid voltage variations, electromagnetic waves occur, the interference signal of which may directly affect the measuring cable and the transient recorder. For this reason the cable and the transient recorder used should be really well shielded. Particularly in the case of an oscilloscope with an amplifier and commercial digital recorders, it is advisable to set them up in a shielded measuring cabin whose line input is fed through an isolating transformer and low-pass filter.

These same aspects which apply for the connection of transient recorders should be observed when direct read-out electronic peak voltage measuring devices are employed. The measurement of impulse voltage chopped on the front is particularly critical.

The peak values of the earth current $\hat{I}_e$ to be expected under unfavourable conditions increase nearly in proportion to the instantaneous value u_d of the chopped voltage. The following guiding values [*Sirait* 1967] have been obtained by experiment:

for incompletely shielded setups $\qquad \hat{I}_e/u_d \le 2.5$ kA / MV

for completely shielded setups $\qquad \hat{I}_e/u_d \le 6.5$ kA / MV

The peak values of i_a can be very different and usually lie well above $\hat{I}_e$. The chopping and the earthing circuit can approximately be interpreted as coupled series resonant circuits; their natural frequencies lie in the region of about 0.5...4 MHz. For a given earth current the impedance of the plane earth conductor is the decisive factor for the voltage drops produced. Further details on the calculation of these impedances can be found in Appendix 4.5. It is possible in this way to estimate the eventual potentials. As a guiding value it is recommended that their amplitudes be restricted to a few hundred volts [*Lührmann* 1973].

2.2.3   Shielding

Extremely sensitive measurements are often performed in high-voltage experiments. Partial discharge measurements in particular can be disturbed when the extended arrangement of the high-voltage circuit behaves as an antenna and receives external electromagnetic waves. Moreover, electromagnetic waves are also produced during breakdown discharge process in high-voltage circuits, and these can in turn effect disturbance of the surroundings.

Practice has shown that the disturbing influence of the surroundings on sensitive high-voltage measurements is generally more intense than that exerted in turn by the high-voltage investigations on the surroundings. This is mainly due to the fact that disturbing pulses occur in high-voltage circuits only occasionally and are short-lived, whereas the external disturbances, due for example to radio-transmitters or improperly screened vehicles or electric motors, generate permanent interference.

Almost complete elimination of external interference and at the same time of eventual environmental influences is achieved by using unbroken metallic shielding in the form of a Faraday cage. The standards required of the plane conductor used for this are appreciably different from those set for the plane conductor in the floor of high-voltage laboratories.

Whereas for conductors intended as return-circuit for the transient current the requirement of a low voltage drop is the primary consideration, high damping of the electromagnetic fields is the objective in plane conductors intended for shielding purposes. It will therefore usually be sufficient if a close-meshed metal net is hung on or set into the walls of the laboratory and the unavoidable apertures for power and communication leads are blocked for high-frequency currents with low-pass filters. When putting this into practice special attention must be paid to careful shielding of doors and windows [*Kaden* 1959; *Prinz* 1965].

The erection of a complete Faraday cage is certainly desirable in every case, but is absolutely necessary only when sensitive partial discharge measurements are intended. Experiments in a high-voltage practical course can usually be performed without exception in partially shielded setups which have only one plane earthed conductor in or on the floor of the laboratory.

2.3 Circuits for High-Voltage Experiments

The electrical circuit for high-voltage experimental setups is appropriately made up of the three circuits shown in Fig. 2.7, namely the power supply circuit 1, the safety circuit 2, and the high-voltage circuit 3.

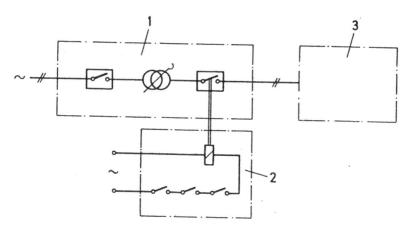

Fig. 2.7 Schematic representation of the basic circuit of a high-voltage experimental setup.
1 Power supply circuit with regulating unit and switchgear
2 Safety and control circuit
3 High-voltage circuit with high-voltage generators, measuring equipment and test object

Besides switchgear equipment, the power supply circuit in most cases contains an element to set the desired voltage. The safety circuit prevents the switching-in, or causes the switching-off, of the high-voltage circuit when one of the safety circuit switches is not closed. Finally, the high-voltage circuit consists of the high-voltage generator and measuring equipment as well as the test object.

2.3.1 Power Supply and Safety Circuits

In addition to the switchgear equipment, the arrangement for setting a variable excitation voltage for the high-voltage source is an important element of the power supply circuit. For power ratings up to 50 kVA, at most 100 kVA, a regulating transformer with carbon rollers is chosen. This is economical and using diverse circuits can be reliably made up to the stated high power ratings.

Installation in the control desk is possible up to about 5 kVA, and manual operation is then most practicable. For higher power, separate installation and remote-controlled motor drive become necessary.

Above about 100 kVA , one has to choose a regulating transformer with metallic contact and load circuit breaker or excitation by means of a synchronous machine. Other possibilities are regulating transformers with rotatable winding. Whilst for excitation via a synchronous generator the supply network is relieved of power impulses during breakdown discharges in the high-voltage circuit, for excitation with regulating transformers these impulses are transferred to the network undamped.

The layout of power supply and safety circuits to be described below corresponds to those used for the practical stands of Fig. 2.1.

The experiments are monitored from control desks which accommodate a regulating transformer for excitation of the testing transformer, as well as the most important measuring and control facilities. As protection against electrical accidents, the safety circuit effects switching-off of the testing transformer on all poles for interruption of any of the series connected safety circuit switches.

The door to the test setup can only be opened with a key which also fits an interlocking switch on the control desk set in the safety circuit. The setup can only be switched on when the door of the practical stand is locked.

A block diagram and the current paths of the setup are reproduced in Fig. 2.8 and 2.9. The control voltage required for energizing the safety and

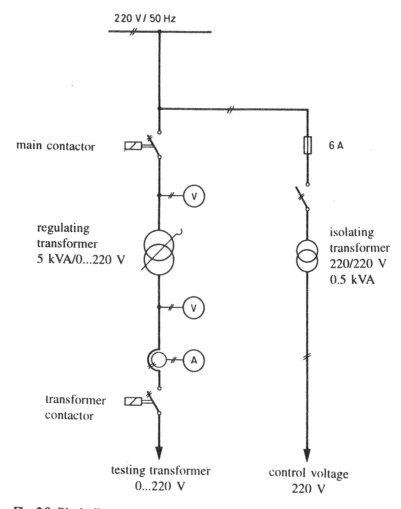

Fig. 2.8 Block diagram for the power supply circuit of a high-voltage experimental setup

control circuits is tapped from the supply voltage via an isolating transformer.

Warning lamps mounted visibly at the entrance of the experimental stand indicate the condition of the transformer switch. Entry to the setup is permissible only when the green warning lamp is on. The entire illumination of the instruments, and of the warning lamps too, can be interrupted for

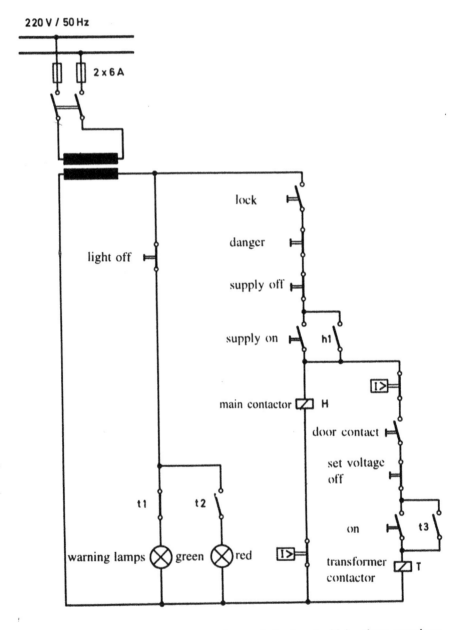

Fig. 2.9 Current paths of the safety and control circuits of a high-voltage experimental setup

short period by the press-button 'light off', to allow observations to be made in total darkness. Set in a particularly prominent position on the control desk is the press-button 'danger', which will immediately release the main switch when pressed. The same holds good when the overcurrent relay indicates overloading of the power supply circuit.

2.3.2   Setting up High-Voltage Circuits

The electrical circuit diagram of high-voltage circuits is usually quite simple, since, apart from the measuring equipment, only comparatively few elements are involved. One particular difficulty, however, is that the specified clearances within the setup and to the surroundings have to be allowed for.

For this reason the electrical circuit diagram alone is not sufficient for the high-voltage circuit; usually it has to be supplemented by a spatial circuit diagram which indicates clearly the three-dimensional arrangement.

The safety clearances given in Appendix 4.1 can also be used as a first estimate of the gap spacings required within a high-voltage circuit. Converted to units convenient for high voltages, we have :

for alternating voltages	5 m for each MV
for direct voltages	3.5 m for each MV
for impulse voltages	2 m for each MV

For switching impulse voltages over 1 MV, data of this kind cannot be given for the required spacings or the necessary minimum clearances. At strongly inhomogeneous electrode configurations, and especially for positive polarity of the electrode with smaller radius of curvature, anomalous flashovers can occur which do not take the shortest path to the opposite electrode but bridge more than double the spacing between two unforeseen points of the electrode configuration. Personnel must be completely protected by metallic barriers during experiments with high switching impulse voltages, since an adequately large clearance alone does not guarantee the necessary safety. Fig. 2.10 shows a flashover for a positive switching impulse voltage of 3.3 MV in an EHV research laboratory. The clearance to the ceiling was 22 m, from which the numerical value of about 7 m per MV could be deduced. But in the same setup, flashovers can take place just as well to both the walls and also to the floor, although clearances between 20 m and 30 m would then be bridged.

Fig. 2.10 Flashover for a positive switching impulse voltage of 3.3 MV in the EHV laboratory of Electricité de France; clearance to ceiling 22 m (Photo H.Baranger, Paris)

As a rule, it is required that no predischarges may occur in a high-voltage circuit up to a certain voltage level. This requirement cannot be fulfilled by adequate clearance alone; the field strength at all points should be kept sufficiently low by appropriate curvature of the metallic parts. Whereas for voltages of at most a few hundred kV it is often still possible to mould the armatures appropriately, separate grading electrodes with large radii of curvature would be needed for very high voltage. These can also be built up of smaller component electrodes, which would considerably simplify their manufacture and transport [*Kind, Kärner* 1982; *Hauschild et al.* 1982].

When setting up high-voltage circuits care should be taken to see that all conductors which could acquire a high potential are earthed before entering the danger-zone. For this purpose safe and easily accessible connecting points for the earth leads should be provided, so that earthing can be done without risk using earthing rods.

In high-voltage generators that contain capacitors, these shall be earthed and short-circuited, when for example, in an impulse voltage generator the resistances are to be interchanged. If work is to be done on the specimen, it is sufficient to earth only the output terminals of the high-voltage generator. High-voltage generators, especially impulse voltage generators and direct voltage setups, have often automatic earthing arrangements.

2.4 Construction Elements for High-Voltage Circuits

Equipment for experiments with high voltage are generally set up in atmospheric air. The dimensions required of the construction elements used depend primarily upon the magnitude of the voltages appearing across them. Apart from this it is necessary to consider the dissipation of operational losses in order to avoid inadmissible overheating.

In the following, several types of the more important high-voltage construction elements for indoor installation shall be briefly described. In the selection of the example, particular priority was given to those applicable in high-voltage practical experiments and also to the feasibility of self-made devices.

2.4.1 High-Voltage Resistors[2]

High-voltage resistors are frequently required as charging resistors, discharge resistors or damping resistors and as measuring resistors too. Here the requirements on accuracy, thermal loading capacity and dielectric strength can be quite different.

Water resistors are especially suitable for applications demanding high thermal loading capacity. Rust-proof electrodes (graphite, stainless steel) are immersed in water which is usually contained in a cylinder or flexible tubing of insulating material. The value of the resistance follows from the length and cross-section of the cylindrical water container and can be varied over a wide range by using additives in distilled or tap water. A specific resistance of about 10^5 Ωcm can be attained with distilled water for longer period; tap water takes values from 10^2 ... 10^3 Ωcm. A stability of better than $\pm 10\%$ can rarely be expected of water resistors. They. are therefore only applicable where moderate demands are made on the accuracy, e.g., as current-limiting resistors in charging circuits. They are also applied in research setups in order to determine the thermal loading limits of test voltage generators. A somewhat higher stability of the resistance value can be achieved with liquid resistors built up of Cu_2SO_4.

---

[2] Comprehensive description in *Marx* 1952; *Craggs, Meek* 1954; *Schwab* 1969.

In one type of construction, suitable for several purposes, a large number of low-voltage resistance elements (wire-wound, layer or compound type resistors) are connected in series. In so doing it is advisable that the individual elements be arranged in such a manner that the external voltage distribution is as uniform as possible. Fig. 2.11 shows two designs as examples, which are suited to high-voltage practicals and correspond in their dimensions and terminal parts to the elements of the high-voltage construction kit described under 2.4.5. In the design shown in Fig. 2.11a, the terminals are arranged between the individual resistance elements so that a voltage divider with finely variable steps is produced. To increase the permissible voltage stress and improve the heat dissipation of each resistance element, they may be immersed in oil as shown in Fig. 2.11b. With respect to the stability of the resistance value, one should take into consideration that the resistance value of high ohmic layer and compound type resistors increases appreciably as a consequence of frequent stressing by rapidly varying voltages [*Minkner* 1969].

This is valid especially for carbon layer resistors which indicate a high temperature coefficient ($500...2000.10^{-6}$ 1/grd) and a strong voltage-dependence during long-term loading. Metal-film and metal-oxide-film are, in spite of the good temperature coefficients ($20...50.10^{-6}$ 1/grd),not so well suited for high-voltage circuits due to their small layer thickness and thereby low self-weight, since the resistance layer can be destroyed by rapid voltage variation (e.g. during a breakdown of the specimen). Best suited for high-voltage circuits are oxide-layer resistors. They possess a selectable temperature coefficient ($40 ... 200.10^{-6}$ 1/grd), and their limited voltage dependence can be made negligible for practical applications by prestressing them e.g. with a number of impulse loadings

In the design of high-voltage resistors, the value of which is independent of load and time, one makes use of metallic conductors made out of e g. Constantan (54% Cu, 45% Ni, Mn) or Nicrothal (75% Ni, 20% Cr, additives). Important is as high a specific resistance as possible (Constantan: 0.5 $\Omega mm^2/m$, Nicrothal: 1.33 $\Omega mm^2/m$) and a temperature coefficient as low as possible (Constantan: 30.10^{-6} 1/grd, Nicrothal: $5 ... 10.10^{-6}$ 1/grd). Due to the relatively low specific resistance, the main problem is the mechanical sensitiveness of the very thin wires required for the high resistance value of, say 10^6 Ω for each kV. For reduction of the construction length, the wires are wound on insulating, very often cylindrical bodies. A coil winding leads to a very high inductance of the resistor, for which reason two layers are often wound in opposite directions and insulated from one another, or better still, built up as a cross-over winding in a

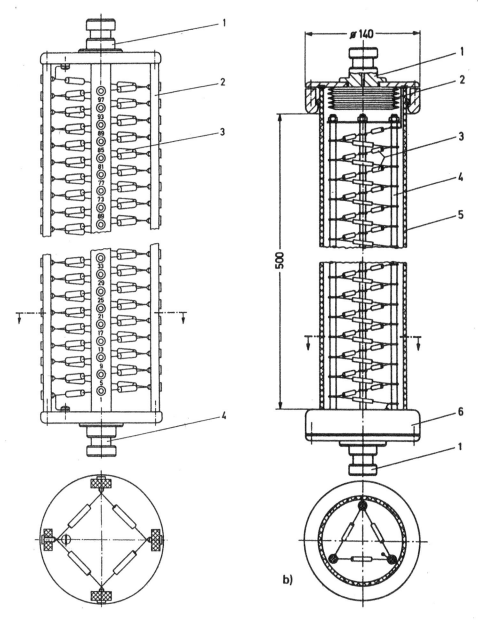

Fig 2.11 2 Types of high-voltage resistors with carbon layer resistors
a) measuring resistor with voltage tap in air, 25 MΩ, 140 kV, continuous duty)
   1 Terminal and fixing bolts, 2 Insulating material   support with plug sockets,
   3 Resistance element
b) load resistor in oil-filled insulating tube, 10 MΩ, 140 kV, short-time duty for
   1 min
   1 Terminal and fixing bolts, 2 Bellows, 3 Resistance elements, 4 Insulating
   material support, 5 Hardboard tube, 6 Metal flange

single layer. In addition, there are also resistance mats, which are wound with meandering glass fibres, which , as low-inductance resistance mats, can be suspended directly in air as resistance bands, or, better still, used in special constructions, e.g. wound on an insulating support and immersed in insulating oil or insulating gas. Their dielectric strength reaches values of the order of 3 kV/cm; resistance values up to 6 MΩ per m² of band area and, with self-cooled surfaces in air, continuous power ratings up to 10 kW can be realized. For particularly stringent conditions, embedding in epoxy resin has proved successful; for high resistance values, mechanically stable and highly stressable resistance elements can also be produced in this way.

Low-inductance resistive dividers for measuring steep impulse voltages are preferred with relatively low resistance values of a few 10 Ω and are built up with compound resistors. In case the low-voltage and the high-voltage part are also built up with the same material, the load-dependent variations in the resistance value compensate each other, and the transformation ratio remains constant.

2.4.2 High-Voltage Capacitors[3]

In addition to resistors, capacitors are the most common elements in high-voltage circuits because they can be manufactured at more or less negligibly low loss for high voltages too. They are employed in circuits for the generation of direct and impulse voltages, as measuring capacitors and also as energy storage devices.

The dielectric of the type of capacitor mostly used consists of several layers of oil-impregnated insulating paper. The thickness of the dielectric is of the order of 50…100 μm. The electrodes are made of thin aluminum foil. A large number of machine-wound capacitors with partial voltages of a few kV are connected in series. For high voltages and high energy contents, the resulting pile of capacitors is usually placed inside a thin sheet metal container which can take the thermal expansion. Load capacitances and measuring capacitances are usually built into an oil-filled cylindrical insulating container; the capacitors CM, CS and CB of the high-voltage construction kit mentioned under 2.4.5 are made in this way. As impregnating medium, synthetic liquids or insulating gases may also be considered besides insulating oil and castor oil (for direct and impulse

---

[3] Comprehensive treatment e.g. in *Liebscher, Held* 1968; *Kind, Kärner* 1982.

voltages). In measuring capacitances, very low-loss plastic foils are often used instead of the paper dielectric.

In general, ceramic dielectric has a decreasing dissipation factor with increasing frequency. It is therefore suitable for the manufacture of high-frequency capacitors for high voltages [*Hecht* 1959]. However, for technological reasons, the possible voltage per element is limited to values of about 10 kV. To reach higher voltages here again a series connection of several elements must be used, which maybe arranged either in air or in oil.

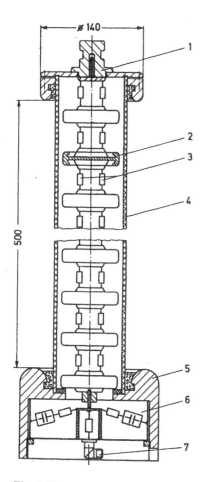

Fig. 2.12 Damped capacitive divider for impulses voltages upto 200 kV built up of ceramic capacitors and layer type resistors in air
1 High-voltage terminal   2 Ceramic capacitor   3 Damping resistors
4 Insulating tube   5 Earthed metal base 6 Low-voltage section
7 Measuring cable terminal

Fig. 2.12 shows as an example of the application of ceramic capacitors, the construction of a damped capacitive impulse voltage divider according to 1.3.13c; its dimensions and terminal parts are also adapted to the high-voltage construction kit described under 2.4.5. One such divider has an extremely short response time of under 10 ns, and can be used for impulse voltages up to 200 kV; the resultant capacitance is 60 pF, the resultant damping resistance 300 Ω. In order to keep the inductance of the low-voltage section as low as possible, a large number of R-C elements is connected in parallel here. The ohmic components of the high-voltage section are also made up of several parallel resistors for the same reason.

Compressed gas is suitable as a dielectric in the production of very low-loss capacitors, such as those needed as reference capacitor C_2 in dissipation factor measuring bridges as per 1.5.3. The arrangement with coaxial cylinder electrodes in particular, after *Schering* and *Vieweg*, has proved to be a practical form of construction. As an example Fig. 2.13 shows a 100 kV compressed gas capacitor developed for the high-voltage practicals and insulated with SF_6 at 350 kPa; its dissipation factor lies below 10^{-5}. This component too was adapted to the dimensions and terminal pats of the high-voltage construction kit described under 2.4.5. For other applications it is most important to provide an appropriate shielding electrode with a high-voltage terminal.

2.4.3   Gaps

Gaps are typical high-voltage construction elements which are used as voltage-dependent or time-dependent switches. The comparatively high resistance of the arc which establishes the conducting path between the electrodes is only rarely a disadvantage in high-voltage circuits. The gap electrodes are usually separated in the non-conducting state by a gaseous medium, preferably atmospheric air, so that repeatability of the switching process is ensured. Gaps in either liquids or solids are used only in rare cases.

Gaps with two electrodes function as voltage-dependent switching devices. They can therefore be employed as protective gaps to prevent excessive overvoltages, as switching gaps in impulse voltage circuits or as measuring gaps for voltage measurement. A few of the most commonly used electrode configurations for 2-electrode gaps are shown in Fig. 2.14.

The plate-plate gap as in a) with Rogowski profile can be used to determine the breakdown voltage in a homogeneous field and is therefore

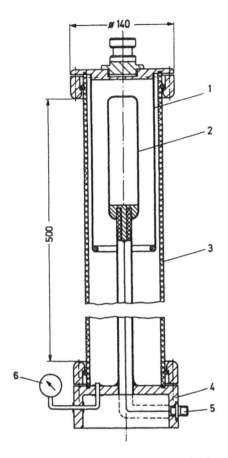

Fig. 2.13 Reference capacitor for 100 kV, 26 pF with compressed gas insulation

1 High-voltage electrode	4 Earthed metal base
2 Measuring electrode	5 Measuring cable terminal
3 Insulating tube	6 Manometer

suitable above all for fundamental physical investigations of the breakdown mechanism.

The transition from a homogeneous to an inhomogeneous field can be achieved with the sphere-sphere configuration as in b) and sphere-plate configuration as in c) by variation of only one parameter, namely the spacing s. For this reason, these configurations also lend themselves to basic physical investigations. Sphere-gaps for the measurement of high voltages have already been discussed in detail under 1.1.10.

The field of the coaxial cylinder gaps as in d) can be calculated very accurately and the effect of edge fields eliminated by a design with guard-

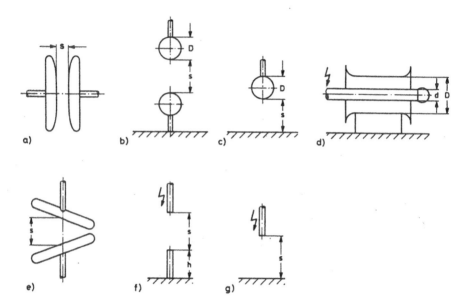

Fig. 2.14 Types of 2-electrode gaps
a) plate-plate (Rogowski profile). b) sphere-sphere. c) sphere-plane, d) coaxial cylinders, e) crossed cylinders, f) rod-rod, g) rod-plane

ring electrodes. Coaxial cylinder gaps are used in particular for investigation
of the incomplete breakdown discharge at wire electrodes; this has great
practical significance with regard to corona discharges on transmission lines.

Crossed cylinders as in e) are also suitable as measuring gaps because
here an almost linear relationship exists between the breakdown voltage
and the spacing for appreciably larger values of s/d than in the case of
spheres.

Rod-gaps represent the prototype of an inhomogeneous configuration.
The electrodes are sharp-edged rods of 100 mm^2 cross-section or
hemispherical in case of round rods (< 20 mm). It has become evident that
the behaviour of the rod-rod configuration as in f) and the rod-plane
configuration as in g) corresponds quite well to that of comparable electrode
arrangements in practical high-voltage laboratories. Rod-gaps are used as
measuring gaps for direct voltages (see 1.2.9). They can also be used for
measuring alternating and lightning impulse voltages at a spacing from
about 300 mm upwards, where the advantage of an approximately linear
relationship between breakdown discharge voltage and spacing follows
[*Wellauer* 1954; *Roth* 1959]. For switching impulse voltages on the other
hand, the configuration positive rod against plate in particular leads to

abnormally low breakdown voltages. By gradual reduction of the height h of the earthed rod in the rod-rod configuration the electrical behaviour changes to that of a rod-plane configuration, and this is of consequence to investigations of the polarity effect and the breakdown discharge mechanism of large air clearances [*Kind, Kärner* 1982].

Gaps with only a weak inhomogeneous field can be extended to function as time-dependent switching devices by the introduction of an auxiliary electrode; these would then, in certain ranges, be independent of the voltage between the main electrodes [e.g. *Deutsch* 1964]. To this end , a voltage pulse is applied between the auxiliary electrode and the surrounding main electrode at the desired instant, and this initiates a flashover between the auxiliary electrode and the surrounding main electrode and thereby triggers discharge between the main electrodes. This type of triggering the switching mechanism is known as cross triggering. This property is exploited in the time-controlled triggering of impulse voltage circuits, in chopping impulse voltages or in the simultaneous triggering of parallel impulse current circuits.

The trigger-range of cross triggering depends upon the polarity of the voltage to be switched and that of the triggering pulse. The more homogeneous a switching gap is, the greater is the trigger- range. The ignition delay time of the switching gap and the spread of the ignition delay time increase with decreasing value of the voltage to be switched, at constant spacing. For a 25 cm sphere-gap, e.g. set up as a switching gap in impulse generators with charging voltage of 200 kV per stage, with a spacing of 8 cm for the first stage, ignition delay times of 100 ns occur nearer to the static breakdown voltage and up to 1 µs at about 25% below the static breakdown voltage. The greatest trigger-range of about 60% is obtained in the case of a positive triggering pulse and a positive voltage to be switched.

As an example, Fig. 2.15 shows the setup of a 3-electrode gap for the high-voltage practical for a maximum working voltage of 140 kV. Installation of the auxiliary electrode is effected in a simple manner by the use of a commercial motor vehicle spark-plug. The trigger pulse, usually fed through a coupling capacitor of about 100 pF, should have a peak value of at least 5 kV.

A simple circuit for a trigger device for controlled triggering of impulse voltage circuits is shown in Fig. 2.16. After firing the thyratron Th a negative voltage pulse appears at the output terminal 1 which is fed to a three-electrode gap either directly or via a delay cable. For triggering the transient recorder, a pulse can be taken from the output terminal 2. Firing the thyratron can be initiated by externally shorting terminal 4, by pressing the

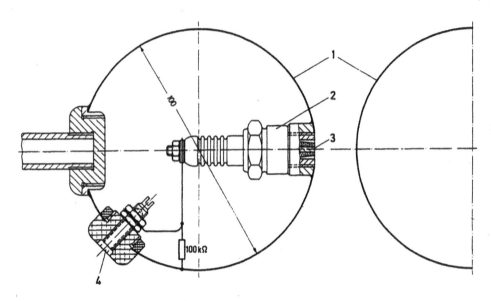

Fig. 2.15 Three-electrode gap for working voltage up to 140 kV.
1 Main electrodes,   2 Spark plug,   3 Trigger electrode, 4 Terminal for trigger pulse

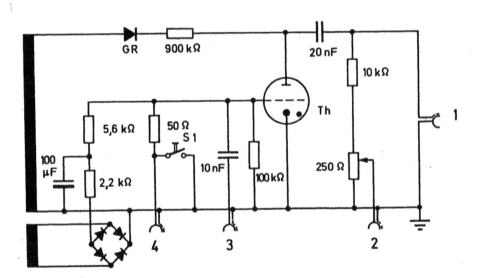

Fig. 2.16 Circuit diagram of a trigger device for controlled triggering of impulse voltage circuits

internal closing contact S1 or by applying a positive voltage pulse to terminal 3.

If parallel connected gaps are to be triggered, instead of the cross-triggering, a longitudinal triggering of the 3-electrode gap must be chosen in order to achieve a negligible spread besides the still shorter ignition delay time. The distances between the three electrodes must be so optimised that the flashover occurs from the trigger electrode first to the main electrode opposite to it. Usually, these gaps are built for a fixed amplitude and polarity of a voltage , so as to discharge a large number of capacitive energy storage systems simultaneously on to a specimen. Thereby the polarity of the trigger pulse, which for a low spread, shall rise as fast and as high as possible (> 20 kV), can also be optimally chosen.

The 3-electrode gap with a static breakdown voltage of 25 kV shown in Fig. 2.17 was developed for simultaneous firing of energy storing capacitors in the voltage range of 15 ... 20 kV. For firing impulse voltages with rate of rise of about 1 kV/ns, the firing time-lag lies below 50 ns [*Petersen* 1965]. These small time-lags have been measured for the given dimensions of the trigger electrode and its insulation. In order to guarantee that these dimensions are unchanged even after a large number of discharges, the material of the trigger electrode e.g. WoCu, as well as its insulation e.g. porcelain, must possess high burn-off resistance.

In switching gaps with trigger electrodes, the trigger range can be set either with the spacing or with the pressure for constant spacing (compressed gas filled gaps).

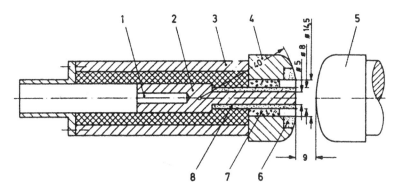

Fig. 2.17 Three-electrode gap for high current capacity and low firing time-lags in the voltage range of 15 to 20 kV.
1 Trigger cable terminal, 2 Trigger electrode,   3 Current circuit terminal,
4 Electrode head, 5 Opposite electrode, 6 Hard metal sleeve (radius of curvature of the bore edge   r = 0.6 mm), 7 Centering for the trigger electrode (PTFE), 8 Ceramic

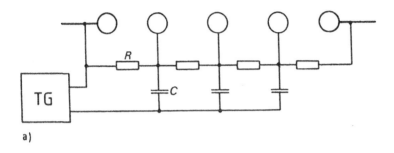

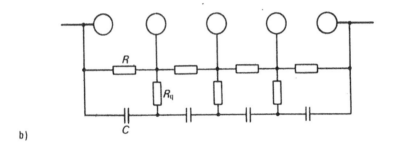

Fig. 2.18 Multiple gaps for high voltages
a) with trigger generator, b) with overvoltage triggering

In special cases, triggering of even simple gaps is effected with the help
of a laser. In addition to a larger trigger-range, a low spread is also realisable
[*Guenther, Bettis* 1978]. The triggering expenses are however extremely
high.

Fig. 2.18 shows the basic circuit diagrams of multiple switching gaps
developed for the switching of higher voltages (> 200 kV) or for chopping
of voltages. The circuit of Fig. 2.18a requires a powerful trigger generator.
The trigger voltage generated from a spiral generator [*Bishop, Feinberg*
1971], for example, is fed via the capacitances and leads to successive
breakdown of the partial gaps. The advantage of this circuit is the larger
trigger-range of about 80%, provided the amplitude of the trigger voltage
is chosen to be the same as the maximum of the voltage to be switched. Its
disadvantage is however the still very high triggering expenses. If this
principle is adopted in a multistage impulse generator, each of the multiple
gaps must contain a trigger generator [*Bishop, Simon* 1972]. The circuit of
Fig. 2.18b has a trigger-range of about 50% only , the triggering expenses
correspond to those of a simple triggering gap in that the first partial gap
is constructed as a 3-electrode arrangement. When these multiple gaps are

introduced in a multistage generator, the multiple gaps of stages 2 to n fire with the help of natural overvoltages that arise on the firing of the first stage [*Feser* 1974]. Multiple gaps with a static breakdown voltage of 200 kV have been realised with upto 20 partial gaps.

2.4.4 High-Voltage Electrodes

The design of high-voltage electrodes and connecting leads is primarily affected by the voltage type and the test assignment. If partial discharge measurements are to be conducted in the test setup, all the electrodes and connecting leads are to be dimensioned such that they are discharge-free. This requirement must be met especially for direct and alternating voltages. No partial discharge measurements are usually conducted with switching impulse and lightning impulse voltages. The requirements here on the high-voltage electrodes for voltage dividers follow from the necessary voltage division and the required measuring accuracy. Discharges at the voltage divider vary the field conditions. But since a change of the earth capacitance affects the transformation ratio, there arises here also the condition of discharge freedom. With switching impulse voltages, the voltage generator must in addition be provided with electrodes, in order to prevent flashovers [*Feser* 1975]. It is only by means of sufficiently large dimensioned electrodes that the leader discharge can be prevented.

A rough estimate of the required dimensions of a spherical top-electrode is obtained from the relationship:

$$r = \frac{U}{E_{max}} .$$

For the maximum permissible field strength E_{max} in air, the electrode roughness, pollution etc. must be taken into account. For indoor arrangements the following permissible field strengths may be assumed for large electrodes [*Feser* 1975]:

for direct voltage:	13 kV/cm
for alternating voltage:	15 kV/cm
for switching impulse voltages:	22 kV/cm
for lightning impulse voltages:	> 30 kV/cm.

With this, e.g. the required radius of a top-electrode for switching impulse voltage of amplitude 1.5 MV works out to 1500/22 = 68 cm.

2.4.5   High-Voltage Construction Kit[4]

In connection with the development of facilities for the High-Voltage Institute of the Technical University of Munich, a system of construction elements was developed by H. Prinz and his co-workers in collaboration with industry. This system allows orderly and rapid setting up of circuits for the high-voltage practical experiments, because of the identity of the external dimensions [Prinz, Zaengl 1960].

Meanwhile, there are various types of high-voltage kits available, in the design of which different points have been emphasised. While the system described below characterises itself by its flexibility and simple construction with elements of the same length and connectivity by the knotting points and appears to be ideally suited for universities, other systems have been developed which are designed for industrial applications (greater energy, firm connections, higher charging voltage per stage).

Naturally, the experiments described in chapter 3 can be conducted with any arbitrary type of high-voltage elements. Nevertheless, the components which were originally designed by Meßwandler-Bau, Bamberg and were preferably used in the high-voltage practicals at the Technical University of Braunschweig, shall be briefly described here as a supplement to the literature.

The dimensions of a basic element for 100 kV alternating voltage or 140 kV direct and impulse voltage are given in Fig. 2.19. It has a hardboard tube casing of 110 mm external diameter; the insulating length is 500 mm. The aluminium flanges at the ends carry a cylindrical terminal bolt with an annular groove for insertion into or mounting on an appropriate support.

Fig. 2.20 shows how a basic element can be connected to auxiliary elements required for mechanical and electrical assembly of a high-voltage circuit. The terminal and fixing bolts for assembling the components of the high-voltage kit can be unscrewed so that the component may be used in another application. The components of the high-voltage kit listed in Table 2.4 are used for the high-voltage practical experiments described in chapter 3.

---

[4] Elements of HV Construction kits are manufactured by:
    Haefely - Trench AG, Basel, Switzerland
    HIGHVOLT GmbH, Dresden, Germany
    MWB India Ltd, Bangalore, India

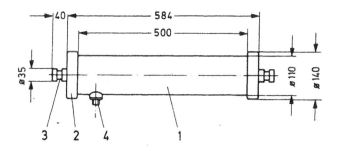

Fig. 2.19 Basic element of the high-voltage construction kit
1 Insulating tube
2 Metal flange
3 Terminal and fixing bolt
4 Measuring terminal for coaxial cable

Fig. 2.20 Auxiliary elements for the mechanical and electrical installation of a basic element of the high-voltage kit
1 Conductive connection rod between the connection cups 2 Connecting cup, 3 Basic element , 4 Floor pedestal, 5 Conductive spacer bar for insertion into the floor pedestal, 6 Insulating support

Table 2.4 Basic elements of the high-voltage kit

Symbol	Construction element	Working data
CM	Measuring capacitor	a.c. 100 kV, 100 pF
RM	Measuring resistor	d.c. 140 kV, 140 MΩ
GR	Rectifier	Peak inverse voltage 140 kV, 5 mA
RS	Protective resistor	Impulse, d.c. 140 kV, 10 MΩ, 60 W
CS	Impulse capacitor	Impulse, d.c. 140 kV, 6 nF
CB	Load capacitor	Impulse, d.c. 140 kV, 1.2 nF
RD	Damping resistor	Impulse 140 kV, 400Ω (1.2/50), 830Ω (1.2/5)
RE	Discharge resistor	Impulse 140 kV, 9500Ω (1.2/50), 485Ω (1.2/5)

The construction elements CM, RM, CB and RE have a measuring terminal attached at the side of the insulating tube, for coaxial cables as shown in Fig. 2.19; this should be shorted when not in use. In addition to the basic elements, other components are available in the high-voltage kit, which have the same terminal dimensions as the basic elements (insulating support IS, sphere-gap KF, conducting link V) or at least conform in their external dimensions to the kit system, e.g. 100 kV testing transformers, measuring gaps, insulated containers for investigations up to 600 kPa, as well as a compressed gas capacitor as a reference capacitor for dissipation factor measurements.

Finally, a Greinacher doubler-circuit according to Fig. 1.22 is shown in Fig. 2.21, first as a three dimensional circuit diagram indicating the spatial arrangement of the individual elements and then as a photograph of the final circuit.

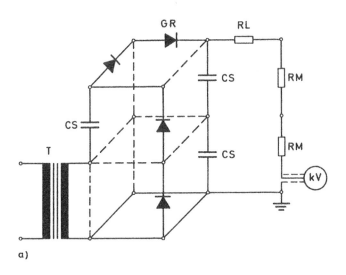

a)

b)

Fig. 2.21 Setup of a Greinacher doubler circuit for 280 kV, 5 mA using the high-voltage kit
a) spatial circuit diagram   b) final circuit (Photo: Meßwandler-Bau, Bamberg)

3 High-Voltage Practicals

In high-voltage practicals, the students are given the opportunity to extend their theoretical knowledge obtained in the lectures by conducting experiments on their own. The experiments can be performed either in small groups of 3 to 6 participants under the guidance of a supervisor, or as joint experiments for a large number of participants. Although group experiments require a large number of experimental setups, thereby imposing a limit on the voltage, they are to be preferred from the point of view of active participation by each member, since this is possible only in small groups. The idea of conducting experiments independently and in a responsible manner is particularly important with regard to the strict observation of safety and operation regulations, essential for the protection of personnel and equipment.

The experiments described here have proved their applicability either in similar or in a slightly modified form, at the High-Voltage Institute of Technical University of Braunschweig, as well as in other educational institutions; they are designed to be carried out in smaller groups with voltages limited to approximately 100 kV. Certain individual phenomena, which make their characteristic appearance only at very high voltages, could also be demonstrated in an additional joint experiment, if need be. The selection and performance of the experiments must at all events be undertaken with the realities of space, technical possibilities and personnel in mind.

The first six of the given experiments correspond to a basic practical which could be recommended as a supplement to lectures for all students of Electrical Engineering in their third academic year. The fundamental principles of experimenting and questions of safety, indispensable to many other experiments, are treated in the experiment "Alternating Voltages". It is therefore strongly recommended that this be conducted first. The sequence of the other experiments can then be chosen arbitrarily.

The next six experiments correspond to an advanced practical, which promotes deeper understanding and experience to those students who desire to specialize in the field of high-voltage techniques. The experiments are concerned with special physical phenomena, with problems of measurement and testing, and with network overvoltages.

Prerequisite for meaningful participation in the practicals is thorough preparation; with each description of the experiment all those sections are mentioned with which one should be familiar, apart from the fundamentals inherent in that particular experiment. While conducting the experiments, the participants should preferably take turns at protocol writing, operation and reading of control and measuring instruments. Evaluation of the results should follow either immediately after the experiments, under the direction of the supervisor, or in a report to be submitted later.

The total duration of an experiment should normally not exceed 3 hours. As an approximate schedule, 30 minutes could be allotted for discussion and confirmation of the necessary prior knowledge, 120 minutes for conducting the experiment proper and 30 ... 60 minutes for evaluation after the experiment. The time spent by the student working on an independent report of the experiment is generally much greater, but the advantages of this method are considerable.

Before beginning the first experiment, each participant of the high-voltage practical must confirm with his signature that he has cognizance of the safety regulations (see Appendix 4.1). In general the experimental circuits are to be set up by the participants, but it is essential that these be checked by the supervisor before each experiment is started. It is strictly forbidden to interfere with the safety circuits of the arrangement !

3.1 Experiment "Alternating Voltages"

Alternating voltages are required for most high-voltage tests. The investigations are performed either directly with this type of voltage, or it is used in circuits for the generation of high direct and impulse voltages.

The topics covered in this experiment fall under the following headings:

- Safety arrangements,
- Testing transformers,
- Peak value measurement,
- RMS value measurement,

- Sphere-gaps.

It is assumed that the reader is familiar with the sections:

- 1.1 Generation and measurement of high alternating voltages,
- 2.3 Circuits for high-voltage experiments,
- Appendix 4.1 Safety regulations.

3.1.1 Fundamentals

a) Equipment of the Experimental Stands
The high-voltage experiments can be set up in experimental stands provided with metal barriers. Control desks with power supply installation, safety circuits and measuring instruments constitute the standard equipment. The control desk circuitry is shown in Fig. 2.8 and Fig. 2.9. For voltage measurement, one meter for measuring the primary voltage of the transformer and one peak voltmeter SM are provided at each desk. The interchangeable low-voltage capacitor of the voltage divider is incorporated in the SM.

b) Methods of Measuring High Alternating Voltages
As shown in 1.1, high alternating voltages can be measured in diverse ways. Of these, the following shall be used in this experiment:

- Determination by using the breakdown voltage $\hat{U}_d$ of a sphere-gap,
- measurement of $\hat{U}$ with the peak voltmeter SM at the low-voltage capacitor of the voltage divider;
- measurement of U_{rms} with an electrostatic instrument;
- determination of $\hat{U}$ using a circuit after *Chubb* and *Fortescue*.

c) Procedure for Determining the Parameters of the Test Voltage Source
As already shown under 1.1.8, the parameters of an alternating voltage source can be determined by three measurements.

3.1.2 Experiment

The following circuit elements are used repeatedly during this experiment:

T   Testing transformer, rated transformation ratio 220 V/100 kV, rated power 5 kVA

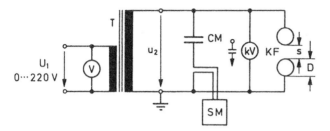

Fig. 3.1 Layout of the test circuit

SM   Peak voltmeter as in Fig. 1.15 with built-in interchangeable low-voltage capacitor. Connection to CM by coaxial cable

KF   Sphere-gap, D = 100 mm

CM   Measuring capacitor, 100 pF

CB   Load capacitor, 1.2 nF

a) Checking the Experimental Setup

The complete circuit diagram of the control desk and the current paths of the safety circuits (examples in Figs. 2.8 and 2.9) should be discussed and, wherever possible, the actual wiring of the experimental setup traced.

A series of measures which guarantee protection against electrical accidents can be identified in the circuit diagram. The faultless functioning of the safety circuit and the fulfillment of the safety regulations of Appendix 4.1 should be checked practically.

b) Voltage Measurements by Diverse Methods

A testing transformer T is connected as shown in Fig. 3.1 single phase to earth. The ratio of the secondary to the rated primary voltage is denoted by $\ddot{u}$; a measuring capacitor CM, a sphere-gap KF and an electrostatic voltmeter are connected on the high-voltage side.

For the gap spacings s = 10, 20, 30, 40 and 50 mm, the breakdown voltage of the sphere-gap should be determined using the following methods:

U_{2rms} by measurement (electrostatic voltmeter, directly or with a capacitive divider)

$\hat{U}_2 / \sqrt{2}$ by measurement (capacitive divider with peak voltmeter SM)

$\hat{U}_d / \sqrt{2}$ from table for sphere gaps, e.g. VDE 0433-2, allowing for air density

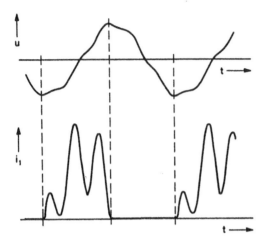

Fig. 3.2 Control oscillogram of the curves of the high-voltage and measuring current in the *Chubb* and *Fortescue* method.

$\hat{U}_2 / \sqrt{2}$ according to the method of *Chubb* and *Fortescue*

For subsequent comparison, the following quantity should also be determined:

$\ddot{u}U_1$ by measurement (moving-coil instrument with rectifier at the control desk)

The surfaces of the spheres should be polished before beginning with the measurements and several breakdowns initiated to remove any dust particles. 5 readings should be taken for each spacing and their arithmetic mean determined.

To determine the peak value by the method of *Chubb* and *Fortescue*, a device with the circuit of Fig. 1.14 replaces SM. The ready-wired low-voltage part comprises two semiconductor diodes D_1 and D_2 as well as a 1 kΩ measuring resistor in the measuring branch for connection of a transient recorder. A moving-coil instrument with 1.5 mA full-scale deflection should be used to measure the current. The curve of the voltage measured with SM should be recorded via the low-voltage capacitor of the capacitive voltage divider and sketched. Similarly, the curve of the current in the measuring branch should also be recorded. Thereby one should check whether the conditions imposed on the curve shape for this method are satisfied, As an example, curves are shown in Fig. 3.2 which can be regarded as just permissible.

Note: Normally, with a testing transformer, no appreciable deviation of the voltage from the sinusoidal form will occur. For demonstration purposes, heavily distorted curves can be generated by connecting an inductance in series with the testing transformer on the low-voltage side. The non-sinusoidal magnetising current then causes a distorted voltage drop across the inductance, which in turn results in the distortion of the input voltage and with that the high-voltage output of the testing transformer.

c) Determination of the Parameters of the Alternating Voltage Source
By switching on, at first the natural frequency of the voltage source without the additional capacitance shall be measured. The measurement shall be made at a voltage of ca. 10 kV. Finally, this measurement is to be repeated with an additional external capacitance (e.g. CB). As a third measurement, a short-circuit shall be created with a sphere-gap at this voltage of 10 to 20 kV (without external capacitance). The curves of the voltage and current before and after the short-circuit are to be measured with a storage oscilloscope or a digital recorder and, from them, the phase displacement between the extrapolated open- circuit voltage and the short-circuit is to be determined.

3.1.3   Evaluation

The breakdown voltage $\hat{U}_d$ of a sphere-gap, determined by the various methods of section 3.1.2b, should be represented in a diagram as a function of s. The origin of the deviation should be qualitatively explained.

Example: Fig. 3.3 shows the required diagram. The measured values were obtained for the comparatively heavily distorted voltage curves shown in Fig. 3.2. The atmospheric conditions were $b = 101.5$ kPa and $T = 296$ K. The tabulated values of breakdown voltage $\hat{U}_{d0}$ according to VDE 0433-2 are

s in mm	10	20	30	40	50
$\hat{U}_{d0}$ in kV	31,7	59	84	105	123

From the individually measured values of $\dfrac{\hat{U}_d}{\sqrt{2}} = \dfrac{d\hat{U}_{d0}}{\sqrt{2}}$, the proportionality factor relating $\hat{U}_2/\sqrt{2}$ and $\bar{I}_1$ for the method of *Chubb* and *Fortescue* should be determined and compared with the theoretical value. The measured

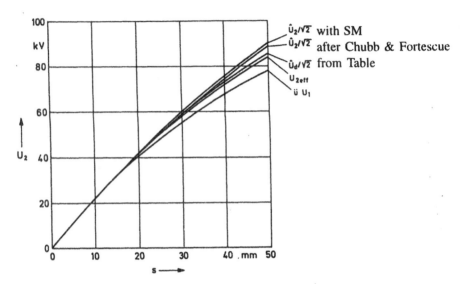

Fig. 3.3 Diagram of voltages measured as per various methods of 3.1.2b

comparative values should also be plotted in the diagram with appropriate characterization.

Based on Fig. 3.2, the time-dependent curve of i_1 should be sketched for the case of a non-smooth measuring voltage u_2. Further it should also be shown why the result would be erroneous if rectifiers were used.

From the natural frequency measurements and the short-circuit measurement as per 3.1.2c, the characteristic parameters R_T, L_T and C_T of the equivalent circuit of the alternating voltage setup should be calculated.

Literature: *Potthoff, Widmann* 1965; *Schwab* 1969; *Stamm, Porzel* 1969; *Kuffel, Zaengl* 1984

3.2 Experiment "Direct Voltages"

High direct voltages are necessary for testing insulation systems, for charging capacitive storage devices and for many other applications in physics and technology. The topics covered in this experiment fall under the following headings:

• Rectifier characteristics,

• Ripple factor,

- Greinacher doubler-circuit,

- Polarity effect,

- Insulating screens.

It is assumed that the reader is familiar with the section

- 1.2: Generation and measurement of high direct voltages.

Note: Extra care is essential in direct voltage experiments, since the high-voltage capacitors in many circuits retain their full voltage, for a long time even after disconnection. Earthing regulations are to be strictly observed. Even unused capacitors can acquire dangerous charges!

3.2.1 Fundamentals

a) Generation of High Direct Voltages
High direct voltages required for testing purposes are mostly produced from high alternating voltages by rectification and, wherever necessary, subsequent multiplication. An important basic circuit for this purpose is the Greinacher doubler-circuit of Fig. 1.22 which can at the same time be considered as the basic unit of the Greinacher cascade. The transient performance of this circuit when switched on can be observed in the voltage curves of Fig. 3.4; after switching the transformer on, the potential of a and b increase in accordance with the capacitive voltage division, since V_2 conducts. At time t_1, V_2 stops conducting and the potential of b remains constant. The potential of a now follows the transformer voltage c, reduced by the constant voltage on C_1, which is indicated by the vertical hatching. At t_2, the diode V_1 prevents the potential of a from falling below zero. Within the time t_2 to t_3 a current flows through V_1 which reverses the charge on C_1. At t_4, voltage division takes place once more and the entire process is repeated until steady-state condition is reached.

If a measuring capacitor is connected to the direct voltage and the alternating current through this capacitor is measured oscillographically, one can determine the ripple $u(t)-\hat{U}$ as in 1.2.11. If a capacitive voltage divider is used, together with a peak voltmeter, its reading would then be proportional to the peak value δU. For low ripple values, the following relationship is valid:

$$\delta U \approx \bar{I_g} \frac{1}{2fC} .$$

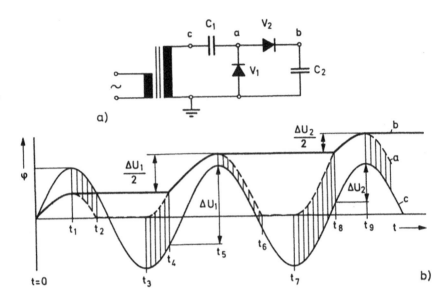

Fig. 3.4 Circuit diagram and voltage curves in a Greinacher doubler-circuit
a) circuit diagram, b) voltage curves for $C_1 = C_2$

b) Polarity Effect in a Point-Plane Gap
At an electrode with strong curvature in air, collision ionisation results
when the onset voltage is exceeded. On account of their high mobility, the
electrons rapidly leave the ionising region of the electric field. The slower
ions build up a positive space charge in front of the point electrode and
change the potential distribution as shown in Fig. 3.5.

When the point electrode is negative, the electrons move towards the
plate. The remaining ions cause very high field strengths immediately at
the tip of the point electrode, whereas the rest of the field region shows
only slight potential differences. This impedes the growth of discharge
channels in the direction of the plate.

For a positive point electrode, the electrons move towards it and the
remaining ions reduce the field strength immediately in front of the point
electrode. Hence, since the field strength in the direction of the plate then
increases, this favours the growth of discharge channels.

*c) Insulating Screens in Strongly Inhomogeneous Electrode Configurations
in Air*
For electrode configurations in a strongly inhomogeneous field, space
charges appear before complete breakdown and their distribution has a

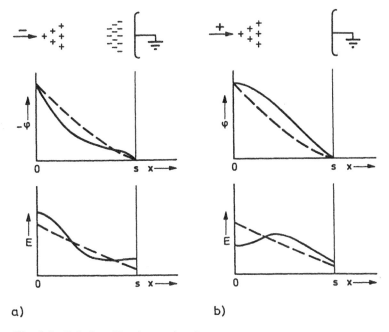

Fig. 3.5 Polarity effect in a point-plane gap
a) negative point, b) positive point

considerable influence on the breakdown voltage. Thin screens of insulating material act as a hindrance to the spreading of these space charges which cause changes in the field strength development. Because of the charge carriers accumulating on the screen, a surface charge with the same polarity as the point electrode results. Thus insulating screens, depending on their position in the field, can cause a variation of the breakdown voltage, which could under certain circumstances, be appreciable.

Fig. 3.6 shows the result of an experiment to illustrate the effect of thin screens in an inhomogeneous field. A photographic plate was arranged along the axis of a point-plane gap. An insulating screen was placed between the electrodes, perpendicular to the photographic plate, at a distance of 9 mm from the point electrode. The figure shows the exposure on the photographic plate after a positive direct voltage of 45 kV had been applied for a few seconds.

If the screen were to be placed directly on one of the electrodes, it would have no effect whatsoever, since the space charge can then either build up without hindrance, or would disrupt the screen immediately. In a homogeneous field, however, a screen has no effect, since no space charges occur.

Fig. 3.6 Discharge photograph showing the field pattern in a point-plane gap with an insulating screen.

3.2.2 Experiment

For this experiment, the following circuit elements will be used repeatedly:

T   Double-pole insulated testing transformer with central tap on the high-voltage winding, rated transformation ratio 220 V/50–100 kV, rated power 5 kVA

GR Selenium rectifier, peak inverse voltage 140 kV, rated current 5 mA

SM Peak voltmeter (see 3.1)

GM D.C. Voltmeter (moving coil ammeter for connecting to RM, 1 mA $\triangleq$ 140 kV, class 0.5)

RM = 140 MΩ, RS = 10 MΩ, CS = 6000 pF, CB = 1200 pF, CM = 100 pF

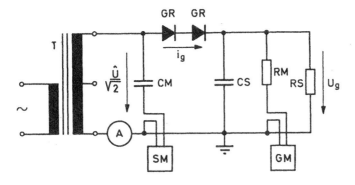

Fig. 3.7 Experimental setup for determining the load characteristic of selenium rectifiers

a) Load Characteristic of Selenium Rectifiers
Using the components mentioned above, the circuit of Fig. 3.7 should be set up. The arithmetic mean value $\bar{I}_g$ of the current through the rectifiers is measured with a moving coil ammeter in the earthing lead of T. The alternating voltage $\hat{U} / \sqrt{2}$ should be set to 50 kV. The amplitude of the direct voltage $\bar{U}$ should be measured for the following cases

- Loading only by the measuring resistor RM ($I_g \approx 0.5$ mA)
- Additional loading by RS ($I_g \approx 5$ mA)

b) Determination of the Ripple Factor
The circuit can now be extended for full-wave rectification according to Fig. 3.8. A transient recorder TR is connected parallel to the peak voltmeter. The direct current $\bar{I}_g$ as well as the peak value of the ripple δU, should be measured with the peak voltmeter SM and observed on the oscilloscope at the same time.

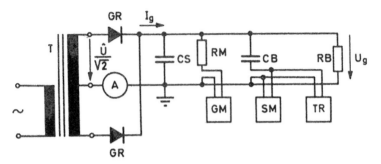

Fig. 3.8 Experimental setup for determining the ripple factor

c) Greinacher Doubler-Circuit
The circuit in Fig. 3.9 should be set up. The variation in potential at point
b with respect to earth is to be recorded . The amplitude of the direct
voltage at *b*, as well as the primary voltage of the transformer, should also
be measured.

d) Polarity Effect
A point-plane gap, in series with a 10 kΩ protective resistance, is connected
in parallel to the measuring resistance RM in the circuit of Fig. 3.9. The
breakdown voltage of this spark gap should be measured for both polarities,
at spacings s = 10, 20,. 30, 40, 60 and 80 mm. The transformer voltage
may not be increased beyond 50 kV in this experiment, to avoid overloading
of the rectifiers and capacitors.

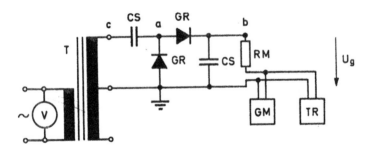

Fig. 3.9 Experimental setup of a Greinacher doubler-circuit

The relationship between breakdown voltage and spacing shown in Fig.
3.10 was obtained for this experiment. One can see that for larger spacings
and a positive point electrode, the excess positive ions in the field region
lead to a lower breakdown voltage.

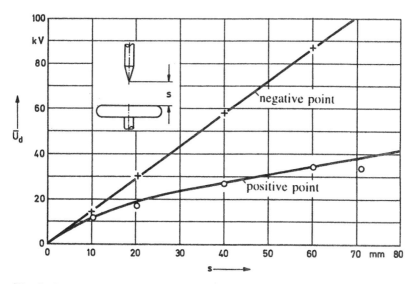

Fig. 3.10 Polarity effect in a point-plane gap

e) Insulating Screens

The setup of d) is retained and the spark gap adjusted to $s = 70$ mm. A paper screen is held between the electrodes perpendicular to their axis, using a device for adjustment (see Fig. 3.6). The breakdown voltage U_d should be measured for positive point electrode, with the screen placed at $x = 0, 10, 20, 40, 60$ and 80 mm.

During these measurements the dependence of breakdown voltage upon the position of the screen, shown in Fig. 3.11 was obtained.

3.2.3 Evaluation

The approximate curve of the load characteristic $U = f(I_g)$ as obtained under a) should be plotted. The number of series-connected rectifier plates for $U_{plate} = 0.6$ V and the value of k should be calculated.

The ripple factor measured under b) should be compared with the calculated value.

In the measurement according to c), how large is the relative deviation of the actual direct voltages from the ideal value, calculated from the primary voltage of the transformer ?

The breakdown voltages for both the polarities measured under d) should be shown graphically as a function of spacing.

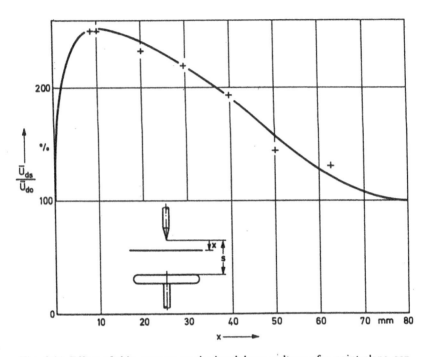

Fig. 3.11 Effect of thin screens on the breakdown voltage of a point-plane gap

The value of $\bar{U}_d$ with screen, referred to the value without, should be represented as a function of x.

Literature: *Marx* 1952; *Lesch* 1959; *Roth* 1959; *Kuffel, Zaengl* 1984

3.3 Experiment "Impulse Voltages"

High-voltage equipment must withstand internal as well as external overvoltages arising in practice. In order to check this requirement, the insulating systems are tested with impulse voltages. The topics covered in this experiment fall under the following headings :

- Lightning impulse voltages

- Single stage impulse voltage circuits

- Peak value measurement with sphere-gaps,

- Breakdown probability.

It is assumed that the reader is familiar with the sections:

- 1.3 Generation and measurement of impulse voltages,

- Appendix 4.6 Statistical evaluation of measured values.

3.3.1 Fundamentals

a) Generation of Impulse Voltages
The identifying time characteristics of impulse voltages are given in Fig. 1.35. In this experiment lightning impulse voltages with a front time $T_1 =$ 1.2 μs and a time to half value $T_2 = 50$ μs are mostly used. This 1.2/50 form is the one commonly chosen for impulse testing purposes.

As a rule, impulse voltages are generated in either of the two basic circuits shown in Fig. 1.36. The relationships between the values of the circuit elements and the characteristic quantities describing the time-dependent curve were given in 1.3.3.

When designing impulse voltage circuits, one should bear in mind that the capacitance of the test object is connected parallel to C_b and hence the front time and the efficiency η in particular can be affected. This has been allowed for in the standards by way of comparatively large tolerances on T_1.

b) Breakdown Time-Lag
The breakdown in gases occurs as a consequence of an avalanche-like growth of the number of gas molecules ionised by collision. In the case of gaps in air, initiation of the discharge is effected by charge carriers which happen to be in a favourable position in the field. If, at the instant when the voltage exceeds the required ionisation onset voltage U_e, a charge carrier is not available at the appropriate place, the discharge initiation is delayed by a time referred to as the statistical time-lag t_s.

Even after initiation of the first electron-avalanche, a certain time elapses, necessary for the development of the discharge channel, which is known as the formative time-lag t_a. The total breakdown time-lag, between over-stepping the value of U_e at time t_1 and the beginning of the voltage collapse at breakdown, comprises these two components, viz.:

$$t_v = t_s + t_a.$$

These relationships are shown in Fig. 3.12.

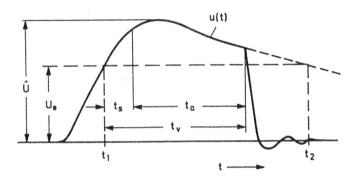

Fig. 3.12 Determination of breakdown time-lag during an impulse voltage breakdown

c) Breakdown Probability

As a condition for breakdown one can roughly expect the time during which the test voltage exceeds U_c (Fig. 3.12) to be greater than the breakdown time-lag t_v. Since t_v is not constant - owing to the statistical scatter in t_s as well as some variation of t_a - repeated stressing of a spark gap with impulse voltages of constant peak amplitude $\hat{U} > U_c$ will not invariably lead to breakdown in every case.

But with each mean value of the breakdown time-lag one can associate an average value of breakdown voltage U_{d-50}, for which half of all the applications result in breakdown.

Thus a breakdown probability P is attributed to each peak value $\hat{U}$ of an impulse voltage of a given form. The distribution function $P(\hat{U})$ is shown in Fig. 1.48 for the case of a sphere-gap. It is zero for $\hat{U} < U_c$ and, in the first instance, reaches a lower limiting value U_{d-0}, referred to as the "impulse withstand voltage" ; knowledge of this is important for designing the insulation levels in installations. U_{d-50} is the value upon which measuring gap applications should be based. The "assured breakdown voltage" U_{d-100} represents the upper limit of the scattering region, which is of significance in protective gaps. Further information on this is contained in VDE 0432-1.

Owing to the asymptotic nature of the distribution function, it is not possible to measure U_{d-0}, and especially U_{d-100}, exactly ; these can, however, be determined with sufficient accuracy if the number of experiments is chosen in accordance with the width of the scattering region. Even in a series with only a few measured values, however, breakdown probabilities can be determined in an approximate manner provided a certain distribution function is assumed. Thus, assuming the Gaussian normal distribution, the following approximation, proved in numerous practical cases, can be used for the arithmetic mean value U_{d-50} as well as the standard deviation s as shown in Appendix 4.6:

$$U_{d-0} \approx U_{d-50} - 3s$$

$$U_{d-100} \approx U_{d-50} + 3s$$

For the evaluation of such an experimental series, the measured values are usually represented on probability paper. If the plot can be approximated by a straight line, a normal distribution may be assumed.

d) Effect of Field Configuration

For a given form of the voltage, the formative time-lag t_a is approximately constant in the homogeneous or only slightly inhomogeneous electric field of a sphere gap. Under stress of about 5% above U_e, t_a is of the order of 0.2 μs. The breakdown probability is therefore determined primarily by the range of the statistical time-lags t_s. This can be greatly minimized by providing for charge carriers in the discharge region, e.g. by UV- irradiation. At low overvoltages, despite irradiation, the mean statistical time-lag can reach values in excess of 1μs. Both t_a and t_s decrease very rapidly with increasing overvoltage $\hat{U}/U_e$.

The spatial as well as the temporal development of a breakdown in an inhomogeneous electric field, as in the case of a point-plane gap or in technical equipment, is different from that in a homogeneous field. Due to spatial restriction of the region in which discharge initiation can occur, the probability of a free charge carrier being there at the instant t_1 is small. The scatter-zone of the breakdown voltage therefore inereases at first with increasing inhomogeneity. By contrast, in such configurations where the onset voltage lies well below the breakdown voltage, charge carriers will be readily available in the electrode vicinity, so that scatter no longer occurs on account of a deficiency of charge carriers whilst the possible breakdown voltage is reached.

In a strongly inhomogeneous field, however, development of the spark channel requires a comparatively longer time than in a homogeneous field; the high charge carrier density must be transferred from the region of highest field intensity to the weaker regions; t_a also increases and is subject to considerable scatter due to the statistical nature of the spatial growth of spark channels.

On the basis of these arguments it can be seen that the breakdown voltage of this kind of configuration, especially for large gaps, varies much more than that of a sphere-gap for instance.

3.3.2 Experiment

The following circuit elements are used repeatedly in this experiment:

T   Testing transformer, rated transformation ratio 220 V/100 kV, rated output 5 kVA

GR   Selenium rectifier, PIV 140 kV, rated current 5 mA

F   Trigger gap, sphere-gap with trigger pin according to Fig. 2.15, $D = 100$ mm

ZG   Trigger generator for generating 5 kV pulses, as in Fig. 2.16

UG   D.C. voltmeter (moving coil ammeter for connection to RM, 1 mA = 140 kV, class 1)

KF   Sphere-gap, $D = 100$ mm.

The data of the construction elements used are as follows:

CS = 6000 pF,   CB = 1200 pF,   RB = 10 MΩ
RM = 140 MΩ,   CM = 100 pF

For impulse voltage 1.2/50 : RD = 416 Ω, RE = 9500 Ω
For impulse voltage 1.2/5 : RD = 830 Ω, RE = 485 Ω

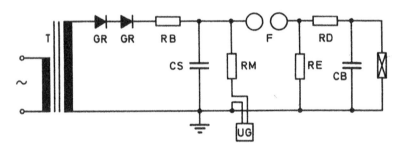

Fig. 3.13 Experimental setup of a single-stage impulse generator

a) Investigation of a Single-Stage Impulse Generator
A single-stage generator is to be set up in circuit b as shown in Fig. 3.13. The voltage efficiency η of the setup should be determined at a d.c. charging voltage U_0 of about 90 kV. The peak value of the impulse voltage $\hat{U}$ should then be measured using the sphere-gap KF. For this purpose, a number of voltage impulses of constant peak amplitude are applied to the sphere-gap, and its spacing is varied until about half the applied impulses result in breakdown. The peak value of the impulse voltage may be determined from the gap length, allowing for the air density. This measurement should

be carried out with both polarities for the 1.2/50 impulse and with negative polarity alone for the 1.2/5 impulse. Using the circuit elements provided for the 1.2/5 impulse, the voltage efficiency for circuit a should also be determined.

This experiment was carried out for circuit b with the voltage impulse form 1.2/50 at relative air density $d = 0.97$, and the following results were obtained:

Charging voltage:	90 kV
Spacing of the sphere-gap:	24.5 mm
$\hat{U}_d$ according to tables:	70.7 kV
$\hat{U}_d$ for d = 0.97:	68.5 kV
η:	81.5%
η calculated from circuit elements:	83.3%

b) Distribution Function of Breakdown Voltage
The single-stage impulse generator should be set up as described in section 3.3.2a using basic circuit b for generation of a positive 1.2/50 lightning surge. The trigger generator ZG, connected as in Fig. 3.14 to the trigger gap F (built up as a three-electrode gap) via the coupling capacitor CM, allows precise triggering of the impulse generator at an accurately preset charging voltage . One of the spheres is provided with a trigger pin to which a voltage pulse is fed through CM. The breakdown between the trigger pin and the surrounding sphere initiates the breakdown of the trigger gap.

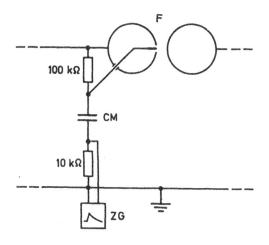

Fig. 3.14 Circuit for triggering a 3-electrode gap

The peak value of impulse voltage is derived using the charging voltage U_0 and the efficiency previously determined in 3.3.2a:

$$\hat{U} = \eta \, U_0.$$

This procedure is permissible here, since the test object capacitance is small compared with the permanently connected load capacitance CB. The chosen test objects are a 10 kV support insulator with protective gap (spacing 86 mm),representative of the inhomogeneous field configuration, and a sphere-gap ($D = 100$ mm, spacing 25 mm) with only a moderately inhomogeneous field. To record the distribution function, the voltage should be increased beyond the breakdown voltage of the test object in steps of about 1 kV, until for 10 impulses initially 0%, finally 100 % flashovers occur.

The measured values of both test objects should be plotted on probability paper and approximated by a normal distribution . Hence the values of U_{d-50} as well as of s, converted to standard conditions, should be determined and the values of U_{d-0} and U_{d-100} obtained approximately. From this experiment the distribution functions $P(\hat{U})$, shown in Fig. 3.15, were obtained. It is evident that the scatter in the breakdown voltages of the

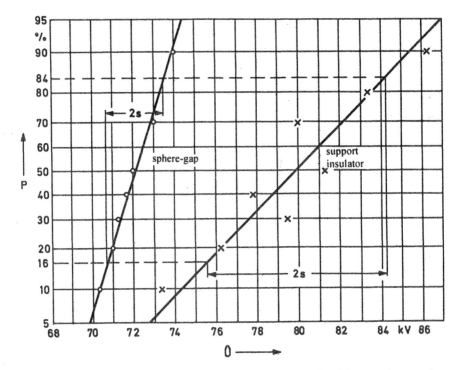

Fig. 3.15 Measured distribution functions of the impulse breakdown voltages of a sphere-gap and a 10 kV support insulator with protective gap

strongly inhomogeneous arrangement of the insulator is appreciably greater than for the case of the sphere-gap.

Evaluation using the straight lines drawn in Fig. 3.15 as approximation for a normal distribution gave:

	Sphere-gap			Support insulator	
U_{d-50}	=	72.2 kV	U_{d-50}	=	80 kV
s	=	1.3 kV	s	=	4.4 kV
$v = s/U_{d-50}$	=	1.8%	$v = s/U_{d-50}$	=	5.5%
U_{d-100}	=	76.1 kV	U_{d-100}	=	93.2 kV
U_{d-0}	=	68.3 kV	U_{d-0}	=	66.8 kV

3.3.3 Evaluation

The characteristic front-time T_1 and time to half-value T_2 should be calculated from the data of the impulse circuit outlined in section 3.3.2a. The measured voltage efficiency should be compared with the theoretical value.

The relationship $P(\hat{U})$ should be determined according to 3.3.2b for the sphere-gap and the support insulator and plotted on probability paper. The values of U_{d-0} and U_{d-100} should be stated for standard conditions.

The scatter ranges of both the arrangements investigated should be compared and reasons given for the differences.

Literature: *Marx* 1952; *Lesch* 1959; *Strigel* 1955; *Kuffel, Zaengl* 1984

3.4 Experiment "Electric Field"

A measure of the electric stress of a dielectric is the electric field strength, the determination of which is therefore an important task of high-voltage technology. The topics covered in this experiment fall under the following headings :

• Graphical field determination,

• Model measurements in electric fields,

• Field measurements at high voltages,

- Numerical field calculation.

It is assumed that the reader is familiar with

- the basic theory of electrostatic fields.

3.4.1 Fundamentals

By the electric strength of an insulating material one understands that value of the field strength which is just permissible under given conditions such as voltage type, stress duration, temperature or electrode curvature. The limit of electric strength of an insulating medium is reached when its disruptive field strength is exceeded at some point. For this reason the determination of the prevailing maximum field strength is of great practical significance.

An exact calculation of the electric field using Maxwell's equations [*Lautz* 1969; *Prinz* 1969], even by applying such special methods as the method of images, conformal mapping or coordinate transformation, is restricted to comparatively few, geometrically simple configurations. Complicated configurations are solved with the help of numerical field calculation techniques. In addition, graphical and experimental methods of determining the electric field have also proved their merit.

a) Graphical Field Determination

The path of the electric filed lines is determined by the direction of the electric field strength $\vec{E}$. They are orthogonal to the equipotential lines at any point and hence perpendicular to metal electrode surfaces. Under the condition that no surface charges exist in the boundary area between two dielectrics, the normal components of the field strengths are inversely proportional to the dielectric constants (DK) of the insulating materials. On the other hand, the tangential component of the electric field strength is continuous along the boundary.

For the case of two-dimensional fields, the field plot can often be obtained graphically with sufficient accuracy. The method is based upon the principle that the equipotential lines and the field lines are estimated first and then the field plot is corrected step-by-step by applying the fundamental electrostatic field laws. Those areas enclosed by adjacent field lines, as in Fig. 3.16, have the same electric flux $\Delta Q = b \, l \, \varepsilon_r \, \varepsilon_0 \, E$, where l is the extension of the configuration perpendicular to the plane of the paper and $\varepsilon_r \, \varepsilon_0 = \varepsilon$ the dielectric constant of the dielectric medium. If the constant

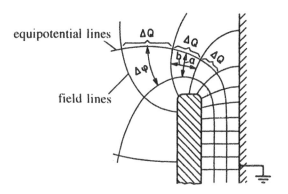

Fig 3.16 Example of a two-dimensional field with field lines and equipotential lines

potential difference between two neighbouring equipotential lines is substituted for E , viz. $\Delta\varphi = E\,a$, the following condition results:

$$\varepsilon_r \frac{b}{a} = k$$

The constant k can be chosen arbitrarily. In the example shown, it is assumed that $b/a = 1$. Let the spacing between two neighbouring equipotential lines be a_1 at any point, then the electric field strength at that point is given by:

$$E_1 = \frac{\Delta\varphi}{a_1}$$

If m is the number of equipotential lines drawn (not counting the electrode surfaces), the total applied voltage is

$$U = (\,m\, +\, 1\,)\,\Delta\varphi.$$

If the number of field lines drawn between the electrodes is n, the total electric flux is given by :

$$Q = n\,b_1\,l\,\varepsilon_0\,\varepsilon_r\,E_1.$$

Substituting, we have for the capacitance of the configuration:

$$C = \frac{Q}{U} = \frac{n}{m+1}\,kl\varepsilon_0$$

A modified version of this method may also be applied to three-dimensional fields, provided these possess rotational symmetry. An analogous argument yields the relationship

$$\varepsilon_r \frac{b}{a} r = k,$$

for the reproduction of the field plot, where r represents the distance of the volume element in question from the rotational axis.

Graphical field determination can be simplified considerably if certain initial data concerning the field configuration are already available. These can be obtained either by calculation or by experimental determination of the field direction, or from the known potentials of individual points. To determine the maximum field strength, it is usually not necessary that the entire field configuration be known, but rather only the field at positions recognised as critical.

b) Analogue Relationships between the Electrostatic Field and the Electric Field

Field measurements on models make use of the analogy between the electrostatic field and the electric field. The following relationships are valid:

Electrostatic field	*Electric field*
$\vec{D} = \varepsilon \vec{E}$	$\vec{S} = \kappa \vec{E}$
$\iint \vec{D} d\vec{A} = Q$	$\iint \vec{S} d\vec{A} = I$

$$\vec{E} = -\mathrm{grad}\,\varphi$$
$$\mathrm{div}\,\mathrm{grad}\,\varphi = 0$$

$C = Q/U$	$1/R = I/U$

The distribution of field lines and equipotential lines follows the same mathematical laws in each case and depends only upon the geometry and materials. Hence the dielectric flux density $\vec{D}$ corresponds to the current density $\vec{S}$ and the dielectric constant ε of the electrostatic field is simulated by the specific conductivity κ of the electric field. If the ohmic resistance R of a configuration is known, the capacitance C can be calculated as:

$$C = \frac{1}{R} \cdot \frac{\varepsilon}{\kappa} \ .$$

These analogue relationships form the basis for the application of the electrolytic tank as well as for electric field simulation with conducting papers.

c) Electrolytic Tank
A scale model of the electrode configuration is set up in a tank with insulating walls, filled with a suitable electrolyte (e.g. tap water). Alternating voltage is the appropriate choice of working voltage, to avoid the polarisation voltages arising in the case of direct voltages. Equipotential lines or equipotential areas in the case of electric field, are measured by means of a probe which can be fed with different voltages from a potential divider via a zero indicator.

Guiding the probe along the lines corresponding to the potential selected on the divider as well as their graphical representation, can be undertaken manually or automatically in large systems. For the two-dimensional field model, various dielectric constants can be simulated by different heights of electrolyte, as shown in Fig. 3.17 for a cylinder-plane configuration. Three-dimensional fields with rotational symmetry can be readily simulated in a wedge-shaped tank, whereas for fields with no rotational symmetry one must resort to much more complicated forms of three-dimensional simulation.

d) Simulation of Electric Fields with Conducting Paper
Two-dimensional fields can be measured easily, and also in most cases with adequate accuracy by means of conducting paper; the number of layers

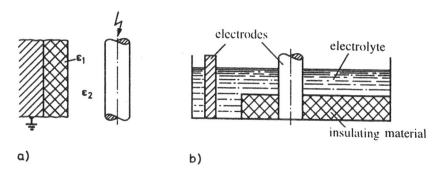

a)   b)

Fig. 3.17 Simulation of a cylinder-plane configuration in the electrolytic tank a) original, b) simulation for the case $\varepsilon_1 = 2\,\varepsilon_2$

of paper arranged one above the other must be chosen to be proportional to the dielectric constant at the respective position. Graphite paper has been found useful as conducting paper, having a surface resistance (viz. resistance measured between opposite sides of a square sample) of about 10 kΩ, similar to that commonly used as conducting layer in high-voltage cables.

The electrode surfaces are represented by conducting silver paint, fixed metal foils, electrically connected needles or spikes driven into a wooden base or by impressed metal objects. At the boundary surfaces between electrodes and dielectrics, or between two different dielectrics, the conducting paper layers must be electrically well-connected with each other. Pins driven into the base are particularly well-suited for this purpose. A distinct advantage of this method is, that the field plot can be directly drawn onto the paper; transfer of the measured values and representation in a separate drawing are thus unnecessary. Basically, this method is also suitable for measuring three-dimensional fields with rotational symmetry. The number of paper layers must then be increased in proportion to the distance from the rotational axis.

e) High-Voltage Field Measurements

The direction of field strength at various points of a configuration in air, as well as the potential of these points can be determined by high-voltage measurements. Photographic record of pre-discharges due to chopped impulse voltages can provide information concerning electrode regions with maximum field strength [*Marx* 1952]. Fig. 3.18 shows this for the example of plate electrodes at a spacing of 200 mm stressed with chopped impulse voltages above 500 kV.

In the primarily practical methods described below, care must be taken to ensure that the measuring cables arranged inside the field zone have only little effect on the electric field. These methods have the definite advantage that they can be applied to manufactured equipment and also take into account the effect of stationary space charges, which appear as a result of the high voltage.

A method for determining the direction of the field strength has been given by *M.Toepler*. It makes use of a small rod-shaped test sample, usually a piece of straw only a few cm long, and is suitable for both direct and alternating voltages. Charges of opposite polarity are induced at the ends of the straw by the field to be measured. This causes a torque which forces the straw, freely suspended at its centre of gravity, to align itself in the direction of the field lines. The position into which this straw test sample

Fig. 3.18 Predischarges in a plate-plate gap during stressing·with a chopped impulse voltage, spacing 200 mm

is deflected corresponds to the direction of the field strength at that point; parallel projection of the straw onto a paper plane parallel to the plane of rotation enables the directions to be marked. If the straw, suspended by an insulation thread is displaced in the same rotational plane, and the traces are joined in accordance with electrostatic field laws, an approximately true reproduction of the field can be constructed.

Three-dimensional spherical field sensors with optical transmission of the measured value can measure the amplitude and direction of the field strength at every point in the field space [*Feser, Pfaff* 1984]. The measuring principle involves the displacement current, so that this type of field measurement is suitable for measuring time variant fields, i.e. for alternating and impulse voltages. The sensor dimensions affect the field to be measured. The sensor must therefore be as small as possible.

Measurement of the potential distribution on the surface of insulating bodies can be carried out with the bridge circuit in Fig. 3.19. The potential on the surface of the test object P is compared here with the known potential tapped off the divider R. A glow lamp of very low capacitance attached to the surface of the test object P is well-suited as a zero indicator S. For

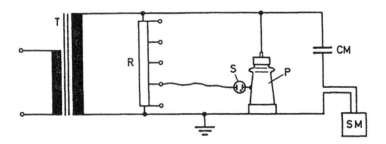

Fig. 3,19 Experimental arrangement for measuring the voltage distribution on the
surface of high-voltage equipment
T High voltage transformer , R Resistance according to Fig. 2.11a,
S Glow lamp of low capacitance, P Test object (porcelain insulating support),
CM Measuring capacitor, SM Peak voltmeter

configurations with rotational symmetry, the capacitive coupling of the glow
lamp can be improved by fixing a wire along an equipotential contour on
the insulating surface.

f) Numerical Field Calculation
A further, predominantly used possibility today is the calculation of the
electric field with the help of numerical methods using computers. Regions
endangered by flashover can be recognised even at the stage of planning a
setup and remedied.

The most important numerical methods for the determination of electric
fields are the method of differences, the finite element method and the
charge simulation method. Other names that could be mentioned are the
surface charge method and the boundary element method.

The method of differences as well as the finite element method necessitate
a discretising of the field space by a network of grid points in which the
desired quantities, e.g. potential or field strength, are determined. The
ensuing matrices, especially in three dimensional arrangements without
symmetry, are often very large; however, in the method of differences,
they are sparsely filled. Open arrangements whose field extends to infinity
are difficult to describe. In the method of differences, the Laplacian
differential equation $\Delta\varphi = 0$ is approximated by a difference statement and
development of the potential φ carried out at each grid point in a Taylor
series, whose coefficients are to be determined from the boundary conditions
i.e. known potentials of the electrodes; whereas, the method of finite
elements in general attempts to minimise the field energy.

The charge simulation method describes the surface charge present at the boundary of an electrode by fictitious, discrete charges in the interior of the electrode. As types of charges, point, line or ring type charges are possible depending on the geometry. The potential φ of the various types of charges can be given by an equation of the form

$$\varphi(\vec{r}) = p(\vec{r})Q \cdot$$

wherein p describes the potential coefficient which is dependent on the location and type of the charge and Q the charge or the charge per unit length. For example, for a point charge at the origin, we have:

$$p(r) = \frac{1}{4\pi\varepsilon r} \cdot$$

At the contour points to be prescribed on the electrode, the potential is known. If the charge location is also specified, then we obtain for the potential φ_k at the contour point k a sum of the N charges,

$$Q_k = \sum_{i=1}^{N} p_k(\vec{r}_k)Q_i$$

which leads to a matrix equation when extended to all the contour points. Finally, the unknown charges can be determined by the inversion of $[p]$:

$$[\varphi] = [p][Q] \cdot$$

The resulting system of equations is clearly smaller than that by the other methods. Charge simulation method can also be applied to open arrangements. Positioning of the contour points and charge points could be automated. For the other methods mentioned, the network could well be generated by a programme and also corrected and refined by hand.

The adequacy of the modelling must be critically verified in every numerical method. That the electrode edge must be an equipotential surface on which the electric field strength is always perpendicular to it could , for example, serve as a criterion.

As a simple example of a plane field problem, an edge-plate arrangement in air shall be considered. The electrostatic field will be calculated using the charge simulation method. Straight, infinitely extended line charges perpendicular to the sketching plane are suitable as the type of charge. Their insertion points are indicated in Fig. 3.20, as also the contour points on the edge of the electrodes. Modelling of sharp bends requires a large number of densely located contour points and charge points close to one another. As a thumb rule, one can mention that each of the four neighbouring

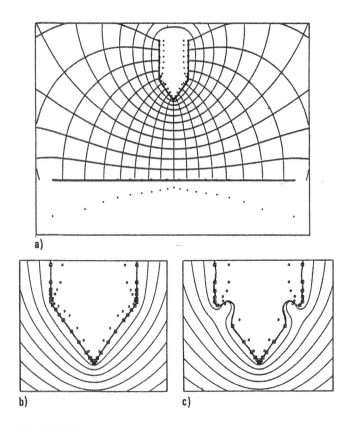

Fig. 3.20 Edge-plate arrangement
a) field lines and equipotential lines of the total arrangement,
b) expected equipotential line along the edge (in detail),
c) insufficient modelling by poor selection of the point positions along the edge (in detail),
charge points •, contour points □

contour points and charge points must form an approximate square. If this rule is disregarded, the calculated field plot as well as the equipotential plot could deviate considerably from those to be expected (Fig. 3.20c).

The calculation resulted in the field plots shown in Fig. 3.20, which show the effect of inadequate modelling.

3.4.2 Experiment

a) Determination of the Equipotential Lines with Conducting Paper
The circuit of the experimental arrangement is shown in Fig. 3.21. The measuring instrument contains a potential divider from which the required

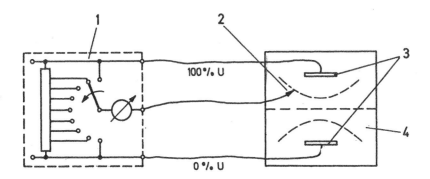

Fig. 3.21 Experimental arrangement for measurement of electric fields with conducting paper
1 Bridge for measuring equipotential lines, 2 Probe, 3 Electrodes, 4 Conducting paper

partial voltages can be tapped, as well as a zero indicator equipped with an amplifier and an indicating instrument. Probe 2 is connected to the corresponding tap on the divider. The measuring voltage is a few volts and the measuring frequency 50 Hz.

A plane electrode configuration in accordance with special instructions should be reproduced on a board with conducting paper, taking the dielectric constants into account. Pins should be inserted at the boundary surfaces between different dielectrics. The electrodes should be connected to corresponding potentials on the potential divider of the measuring instrument. The path of the equipotential lines should be plotted and then drawn.

Fig. 3.22 shows the setup during measurement ; as an example of a configuration with a two-dimensional field, the base of a current transformer with rated voltage $U_n = 20 / \sqrt{3}$ kV is represented. The equipotential lines measured for the same example are shown in Fig. 3.23.

b) Measurement of Fields at High-Voltage

The circuit shown in Fig. 3.19 should be setup. T is a testing transformer for at least 30 kV, CM is a measuring capacitor, SM is a peak voltmeter as in 3.1. As a resistive potential divider R the element shown in Fig. 2.11a, for example, is suitable. The glow lamp S can be fastened to the surface of the test object with wax. As test object P a 30 kV insulating support is used. In a trial series, the field plot is obtained using the straw method under an alternating voltage corresponding approximately to the rated voltage of the test object.

Fig. 3.22 Measurement with the equipotential line measuring bridge

In the second test-series, the glow lamp S is fixed to the surface of the test object and connected to a tap on R. During the actual experiment it should be observed that the measuring accuracy depends upon the ratio of the given glow lamp ignition voltage to the total voltage. The total voltage should therefore be at least 20 times the ignition voltage of the glow lamp. The potential of the probe then corresponds to the mean potential selected on R, to within about 5% of the total voltage.

The traces of the individual positions of the straw shown in Fig. 3.24 were obtained in this experiment. In addition, the 25, 50 and 75% equipotential lines were constructed using some points determined by the glow lamp method and the orthogonality condition for field and equipotential lines.

3.4.3   Evaluation

The points of maximum tangential field strength E_t along the insulator / air boundary should be determined in the equipotential plot obtained with

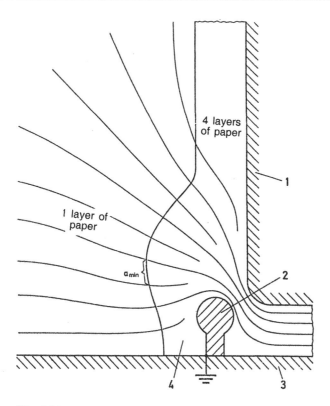

4 layers
of paper

1

I layer of
paper

a_{min}

2

4

3

Fig. 3.23 Equipotential lines of a potential transformer base obtained by simulation
with conducting paper
1 High-voltage winding,   2 Control electrode   3 Earthed plate,   4 Insulation (epoxy
resin)

conducting paper according to 3.4.2b. From the example in Fig. 3.22, the
minimum spacing a_{min} between neighbouring equipotential surfaces is found
to be 9 mm. At the rated voltage $U_n = 20 / \sqrt{3}$ kV, the value of E_t regarded
as permissible is given by:

$$E_t = \frac{0.1 U_n}{a_{min}} = 1.2 \text{ kV} / \text{cm}.$$

Using the results obtained in 3.4.2b by the straw and glow lamp methods,
some equipotential lines should be sketched to a scale of 1 : 1 in the
investigated half-plane.

**Where do the maximum field strengths occur in the edge-plate
configuration?**

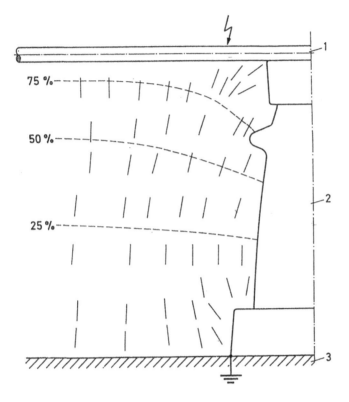

Fig. 3.24 Result of measurement of the electric field by the straw and glow lamp methods
1 Reproduction of a busbar   2 Support insulator   3 Earthed base plate

What types of apparent charges are suited for calculation of arrangements with rotational symmetry e.g. point-plane gap and where should they be arranged?

Literature : *Strigel* 1949; *Marx* 1952; *Küpfmüller* 1965; *Potthoff*, *Widmann* 1965 ; *Philippow* 1966; *Kuffel, Zaengl* 1984; *Beyer et al.* 1986

3.5 Experiment "Liquid and Solid Insulating Materials"

Insulation arrangements for high voltages usually contain liquid or solid insulating materials whose breakdown strength is many times that of atmospheric air. For practical application of these materials not only their physical properties but also their technological and constructional features must be taken into account. The topics discussed in this experiment fall under the following headings :

- Insulating oil and solid insulating material,
- Conductivity measurement,
- Dissipation factor measurement,
- Fibre-bridge breakdown,
- Thermal breakdown,
- Breakdown test.

Familiarity with the following sections will be assumed :
- 1.1 Generation and measurement of high alternating voltages,
- 1.5 Non-destructive high-voltage tests (excluding 1.5.4)

3.5.1 Fundamentals

a) Measurement of the Conductivity of Insulating Oil
The specific conductivity κ of an insulating oil depends strongly on the field strength, temperature and contamination. It is a result of ionic movement and varies in order of magnitude from 10^{-15} to 10^{-13} S/cm for water content ranging from $10...200$ ppm[1]. The measurement of κ yields valuable information about the degree of purity of an insulating liquid. The positive and negative ions are produced on dissociation of electrolytic contaminations. For a specific type of ion with charge q_1 and density n_1, the corresponding contribution to the current density at not too high a field strength $\vec{E}$ is given by:

$$\vec{S}_1 = q_1\, n_1\, \vec{v}_1 = q_1\, n_1\, b_1\, \vec{E}$$

where $\vec{v}_1$ is the velocity and b_1 the mobility of the ions, the latter being constant only when Ohm's law is valid. The corresponding contribution to the conductivity follows :

$$\kappa_1 = q_1 n_1 b_1$$

When a certain field strength is established in the dielectric, a compensating mechanism is set in motion to balance the density of the various types of ions and continues until an equilibrium is established between generation, recombination and leakage of ions to the electrodes. Owing to their different mobilities, this compensating mechanism will be realised at different rates for the diverse ions, which is the reason why the

---

[1] ppm = *parts per million* $\triangleq 10^{-6}$

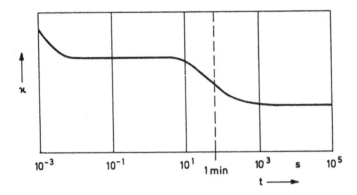

Fig.3.25 Basic dependence of d.c. conductivity of an insulating oil upon the measuring time

resulting conductivity κ is a function of the time after switching on. Fig. 3.25 shows the basic characteristic. For measurement of κ it is therefore advisable to wait until these transient mechanisms have passed and begin with the measurement at a certain time, e.g. 1 min, after applying the voltage.

An electrode arrangement which is to be used to measure κ must be fitted with a guard ring electrode as shown in Fig. 1.66. The electric field should be as homogeneous as possible. Apart from plate electrodes, coaxial cylinder electrodes are commonly used (VDE 0303 and VDE 0370). If the measuring voltage applied is U for a homogeneous field of area A and spacing s, κ is derived from the current I :

$$\kappa = \frac{I}{U} \cdot \frac{s}{A} \quad .$$

The currents to be measured are usually of the order of picoamperes. Sensitive moving-coil mirror galvanometers may be used for this purpose. Current measuring devices with electronic amplifiers are easier to handle, and much more sensitive.

b) Dissipation Factor Measurements of Insulating Oil
Dielectric losses at alternating voltage are due to ionic conduction and polarisation losses. The magnitude and nature of these losses, as functions of temperature, frequency and voltage, are a measure of the quality of the insulation concerned. They also provide information about the physical mechanisms and permit assessment of the suitability of the insulator for particular applications.

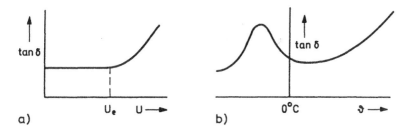

Fig.3.26 Basic dependence of the dissipation factor of an insulation upon voltage and temperature
a) tan $\cdot \delta = f(U)$, b) tan $\delta = f(\vartheta)$

Fig. 3.26 shows examples of the dependence of the dissipation factor tan δ of insulation upon voltage U and temperature ϑ, both of which are of great importance in high-voltage technology. From the rise of the function tan $\delta = f(U)$ at the onset voltage U_e, it can be inferred that partial discharges are initiated either on or within the test sample, causing additional ionic loss. The same shape could also be a result of field strength-dependent variations in the electrolytic conductivity [*Kieback* 1969]. The shape of the function tan $\delta = f(\vartheta)$ indicates the temperature above which loss due to ionic conduction exceeds that due to polarisation.

By definition, the dielectric losses of an insulation with capacitance C at the angular frequency ω can be calculated using the dissipation factor :

$$P_{diel} = U^2 \omega C \tan \delta.$$

It can be measured in a bridge circuit according to Fig. 1.68 or Fig. 1.69, which at the same time allows an exact measurement of the test object capacitance, if the capacitance C_2 of the loss-free standard capacitor is known.

To determine the dissipation factor of liquid or solid materials, basically the same electrode arrangement as used for the measurement of the d.c. conductivity is suitable. The Schering bridge under balanced conditions enables direct reading of the dissipation factor. If the dielectric constant is to be determined, the capacitance C_L of the arrangement in air will also have to be measured in addition to C, so that

$$\varepsilon_r = \frac{C}{C_L} \cdot$$

The relationship $\varepsilon_r = f(U)$ or $\varepsilon_r = f(\vartheta)$ provides supplementary information concerning the physical mechanisms within the insulating material.

c) Fibre-Bridge Breakdown in Insulating Oil

Every technical liquid insulating material contains macroscopic contaminants in the form of fibrous elements of cellulose, cotton, etc. Particularly when these elements have absorbed moisture from the insulating liquid, forces act upon them, moving them to the region of higher field strength as well as aligning them in the direction of $\vec{E}$. The physical explanation for the alignment of the fibrous elements is the same as that given for the straw method in 3.4.1e.

In this way, fibre-bridges come into existence. A conducting channel is created which can be heated due to the resistance loss to such an extent that the moisture contained in the elements evaporates. The breakdown which then sets in at comparatively low voltages, can be described as local thermal breakdown at a defect.

The mechanism is of such great technical significance that in electrode arrangements for high voltages, pure oil sections have to be avoided. This is achieved by introducing insulating screens perpendicular to the direction of the field strength. In the extreme case, consistent application of this principle leads to oil-impregnated paper insulation, which is the most important and very highly stressable dielectric for cables, capacitors and transformers.

d) Thermal Breakdown of Solid Insulating Materials

In solid insulating materials, thermal breakdown can be either total, i.e. a consequence of collective overheating of the insulation, or local, i.e. a consequence of overheating at a single defect. It can be explained by the temperature dependence of the dielectric losses; their increase can exceed the rise in the heat being conducted away, P_{ab}, and so can initiate thermal destruction of the dielectric.

Fig. 3.27 shows the curves of the power P_{diel} fed in at different voltages and the power P_{ab} which can be led away from the test object, as functions of the temperature ϑ which is assumed constant throughout the entire dielectric. Thermal breakdown then occurs when no stable point of intersection for the curves of the input and output power exists. Point A represents a stable working condition and point B, on the other hand, is unstable. If the voltage is increased at a constant ambient temperature ϑ_u, both points of intersection move closer together until, at $U = U_k$, they coincide in C. This voltage is referred to as the critical voltage; at or above U_k, a stable condition is impossible.

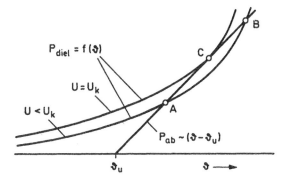

Fig. 3.27 Illustration of the thermal breakdown of solid insulating materials

An increase in $\tan\delta$ at constant voltage indicates that U_k has been overstepped at total thermal breakdown. U_k can therefore be experimentally identified without destroying the insulating material. For inhomogeneous field configurations, one should note that the specific dielectric loss P'_{diel} depends upon the square of E:

$$P'_{diel} = E^2 \; \omega\varepsilon_0\varepsilon_r \tan \; \delta.$$

In regions of maximum field strength, the risk of thermal breakdown is thus particularly high. But, this can be established in dissipation factor measurements only when the dielectric losses in the endangered area, increasing on account of continued overheating, are independently measurable, i.e. can be isolated from the total dielectric losses.

e) Breakdown Strength of Solid Insulating Materials

The experimentally determined values of the breakdown strength of a solid insulating material, owing to the many possible breakdown mechanisms, strongly depend upon the electrode configuration in which they have been measured. A particular problem is the fact that the solid insulating material generally has an appreciably higher breakdown strength than the materials in the vicinity of the testing arrangement, so that there is the risk of a flashover. Some simple test configurations are shown in Fig. 3.28.

The arrangement a) where two plate electrodes are applied to a plane solid insulating material, is restricted in its application to very thin insulating foils, a fraction of a mm thick. This is because, for larger thicknesses, higher voltages would be needed for breakdown, which would lead to gliding discharges at the electrode edges. The breakdown which sets in at these

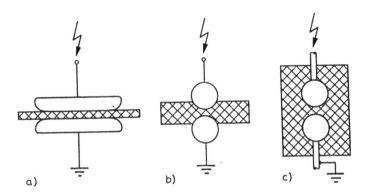

Fig. 3.28 Practical test object configurations for breakdown investigations of solid insulating materials
a) plate electrodes applied to a foil type specimen
b) spherical electrodes inserted in a plate type specimen
c) spherical electrodes cast in an epoxy resin specimen

locations is more characteristic of the electrode configuration rather than of the dielectric.

An increase in the onset voltage can be gained by immersing the arrangement in an insulating liquid. The onset of interfering gliding discharges can be prevented only when the product of the dielectric constant and breakdown field strength for the immersing medium is greater than that for the solid insulating material to be investigated. The setup can in general be used for breakdown voltages of some 10 kV only.

The onset voltage for gliding discharges at the electrode edges can be raised for plate-shaped solid samples by the insertion of spherical electrodes, either on one or both sides of the sample. Through additional immersion in a liquid insulating material, e.g. insulating oil, this arrangement as in b) can be used upto about 100 kV.

Plastics have very high breakdown strengths and are even used as homogeneous insulation at working voltages of the order of 100 kV. A suitable testing arrangement for epoxy resins is the arrangement c) in which two spherical electrodes are cast into a homogeneous block of insulating material. Additional immersion in a liquid insulating material allows breakdown investigations up to some hundreds of kV to be carried out with this arrangement.

For all arrangements, the advantageous effect of immersion in a liquid insulating material can be improved further, since the breakdown strength of the latter increases with the application of higher pressures [*Marx*

1952].The relevant specifications (VDE 0303-2) contain further details for performing the actual breakdown strength measurement.

3.5.2 Experiment

a) Measurement of the D.C. Conductivity of Transformer Oil
In the circuit shown in Fig. 3.29a a sensitive ammeter, with a built-in d.c. voltage source G^2 is connected to an oil-filled testing vessel P. A direct voltage $\bar{U}$ of about 100 V is supplied by the voltage source.

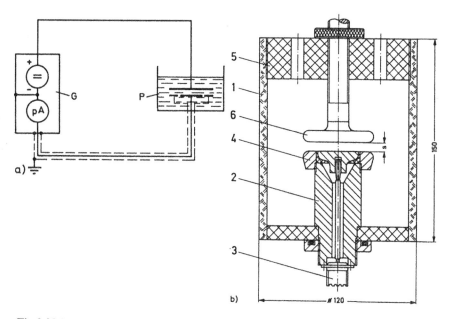

Fig.3.29 Measurement of the d.c. conductivity of insulating oil
a) circuit diagram,   b) testing vessel

The testing vessel P as in Fig. 3.29b consists of a perspex tube 1 with a base of insulating material. A brass bolt 2 with central bore and coaxial socket 3 are fitted through an opening in the base. The brass bolt supports a guard-ring electrode 4. The testing vessel can be closed with a lid 5; the upper electrode 6 is attached to a threaded shaft. The gap *s* can be adjusted by turning the threaded shaft in the lid.

---

[2] Picoammeter manufactured by M/s. Knick, Berlin.

The d.c. conductivity of two different oils should be measured with gap spacings s in the range of 2...5 mm. Fig. 3.30 shows the measured curves for the conductivity of two oils as a function of the measuring time; whereas the conductivity of one oil remains constant throughout the measurement, that of the other oil decreases continuously.

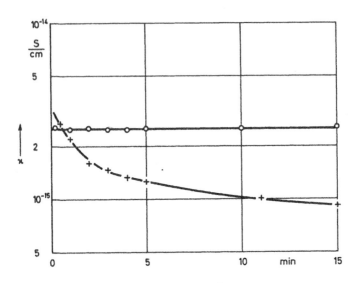

Fig.3.30 Specific conductivity of two transformer oils as a function of measuring time

b) Measurement of the Dissipation Factor of Transformer Oil

The capacitance C_x and the dissipation factor $\tan\delta$ of the arrangement specified in 3.26a should be measured as a function of the a.c. test voltage U, using the circuit shown in Fig. 3.31. The voltage generated by the high-voltage transformer T is measured using the measuring capacitor CM and the peak voltmeter SM. In parallel with the testing vessel is the standard capacitor with capacitance $C_2 = 28$ pF (compressed gas capacitor as in Fig. 2.13).

A series of measurements at voltages up to 35 kV and spacing $s = 5$ mm on a repeatedly pre-stressed oil sample yielded the curve $\tan\delta = f(U)$ shown in Fig. 3.32. The curve, a result of regular measurements at increasing and decreasing testing voltage, shows distinct hysteresis. Both branches of the curve, however, increase continuously with the voltage. From about 27 kV onwards, the influence of the measuring time on the dissipation factor becomes noticeable. During each measurement, the applied voltage was held constant for 2 min. The values measured at the beginning of each

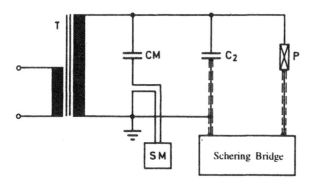

Fig.3.31 Circuit for measuring the capacitance and dissipation factor of a test object with the Schering bridge

such interval were lower than at the end; mean values have been plotted in the diagram.

c) Fibre-Bridge Breakdown in Insulating Oil
In the setup shown in Fig. 3.29b, the electrode is replaced by a sphere e.g. of 20 mm diameter, and the spacing set to a few cm. Some slightly

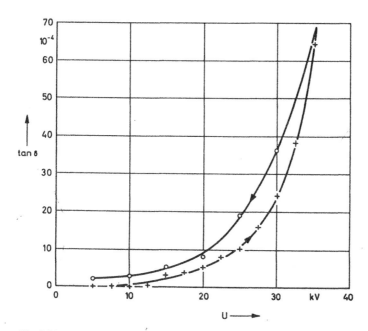

Fig.3.32 Dependence of the dissipation factor of a transformer oil upon test voltage

moistened black threads of cotton 5 mm long are contained in the oil. A voltage of about 10 kV applied between the sphere and plate, within a few seconds, results in the alignment of the threads in the direction of the field ; a fibre-bridge is established, which can either initiate or accelerate a breakdown. The two photographs of the model experiment shown in

Fig. 3.33 Model experiment showing fibre-bridge formation in insulating oil
a) fibres before switching the voltage on
b) fibre-bridge 1 minute after switching the voltage on

Fig. 3.33 indicate clearly the extent to which oil gaps in high-voltage apparatus, which are not subdivided, are exposed to risk by dissociation products and other solid particles.

d) Breakdown of Hardboard Plates

In accordance with VDE 0303-2, the 1-minute withstand voltage of 1 mm thick plates of a hardboard sample should be determined as follows. The test circuit is the same as in 3.5.2c. The breakdown voltage should be determined approximately in two preliminary trial runs with a rate of voltage rise of 2....3 kV/s. The resulting mean value, as breakdown voltage U_{dm} will be taken as the basis for future experiments. In the first minute of stressing, a voltage 0.4 U_{dm} should be applied. Then the voltage should be increased by 0.08 U_{dm}, again held for 1 min, and so on, until breakdown occurs. The voltage at which the insulation was just on the verge of breakdown is the 1-minute withstand voltage. The 5-minute withstand voltage should be determined in a similar way; as a rule, it is appreciably lower.

3.5.3 Evaluation

The time-dependent characteristic of the d.c. conductivity of transformer oils, measured as in 3.5.2a, should be graphically represented.

The dissipation factor and the capacitance of the oil-filled testing arrangement, measured as in 3.5.2b, should also be shown graphically as a function of the voltage. The formation of fibre-bridges should be noted in the testing arrangement as in 3.5.2c.

The 1-minute and the 5-minute withstand voltages of 3 plates of a 1 mm thick hardboard sample should be determined according to the method given in 3.5.2d. The ratio of the two withstand voltages should be calculated from the mean values and discussed.

Literature : *Whitehead* 1951; *Böning* 1955; *Imhof* 1957; *Lesch* 1959; *Roth* 1959; *Anderson* 1964; *Potthoff, Widmann* 1965; *Kind, Kärner* 1982

3.6 Experiment "Partial Discharges"

In insulation systems with strongly inhomogeneous field configurations or with an inhomogenous dielectric, the breakdown field strength can be locally exceeded without complete breakdown occurring within a short time. Under these conditions of incomplete breakdown, the insulation between the electrodes is only partially bridged by discharges,. These partial discharges (PD) have considerable practical significance, particularly for the case of stress by alternating voltages.

The topics covered in this experiment fall under the following headings:

- External partial discharges (Corona),

- Internal partial discharges,

- Gliding discharges.

It is assumed the reader is familiar with section

- 1.5 Non-destructive high-voltage tests.

3.6.1   Fundamentals

In a strongly inhomogenous field, external partial discharges occur at
electrodes of small radius when a definite voltage is exceeded. These are
referred to as corona discharges and, depending upon the voltage amplitude,
they result in a larger or smaller number of charge pulses of very short
duration. It is these discharges which are the source of the economically
significant corona losses in high-voltage overhead lines; moreover, the
electromagnetic waves generated by the charge pulses can also cause radio
interference.

Partial discharges can also occur inside high-voltage equipmentat a
distance from the electrode surfaces, particularly in gas inclusions in solid
or liquid insulating materials (cavities, gas bubbles). Hence there is the
risk of damage to the dielectric as a result of these internal partial discharges
during continuous stress, due to breakdown channels developing from such
partial discharge sites and because of additional heating.

Partial discharges which develop at the interface of two dielectrics in
different states of aggregation are known as gliding discharges. Especially
when the interface under stress is in close capacitive coupling with one of
the electrodes, high energy discharges take place which, even at moderate
voltages, can bridge large insulation lengths and so damage the insulating
materials.

a) Partial Discharges at a Needle Electrode in Air
The most important physical phenomena of external PD at alternating
voltage can be observed particularly well on the example of a needle- plate
electrode configuration in air. Fig. 3.34 shows a suitable configuration.

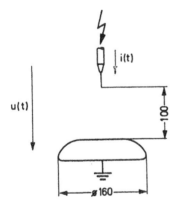

Fig. 3.34 Needle-plate gap

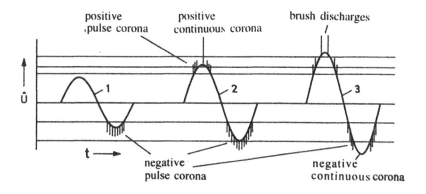

Fig. 3.35 Types of appearance and phase dispostion of partial discharges

As curve 1 in Fig. 3.35 schematically shows, when the applied voltage is increased, pulses first appear at the peak of the negative half-period; the amplitude, form and periodic spacing of these are practically constant. These are the so-called " Trichel Pulses", also observed under negative direct voltages and on the evidence of which *G.W.Trichel* in 1938 demonstrated the pulse-type character of corona discharges. The pulse duration is a few tens of ns and their frequency can be up to 10^5 s^{-1} If the voltage is increased further, pulses also appear at the peak of the positive half-period, however, these are irregular (curve 2).

For both polarities, with increasing voltage, in the peak region, pulse-less partial discharges may also occur, referred to as "continuous corona" (curves 2 and 3) ; this is the reason for the comparatively low radio interference found in some cases, despite extensive corona losses. The final typical discharge mode prior to complete breakdown is intense brush discharges in the positive peak (curve 3).

The pulse-type character of the pre-discharges may be explained using the Trichel-pulse example. The electron avalanches produced at the negative point electrode travel in the direction of the plate. Their velocity is strongly reduced owing to the rapidly decreasing field strength (see Fig. 3.5), and by attachment of electrons to the gas molecules, negative ions are formed. The space charge so produced reduces the field strength at the cathode tip, thus preventing further formation of electron avalanches. A new electron avalanche can commence from the cathode only after removal of the space charge by recombination and diffusion. The pulse-type discharges occur in the region of the test voltage peak.

b) Corona Discharges in a Coaxial Cylindrical Field

Corona performance of overhead line conductors is of great significance to the technical properties and economics of a high-voltage line. Corona measurements can be carried out in the laboratory, if the conductor arrangement to be studied is chosen to be the inner electrode of an assembly of coaxial cylinders. In such a "corona-cage", the field configuration near the conductor differs very little from that of the actual transmission line, since one may safely assume that the conductor spacing in the latter is very large compared with the conductor radius and therefore the field in the vicinity of the conductor similarly possesses cylindrical symmetry.

Fig. 3.36 shows an arrangement which can be used for a.c. experiments up to about 80 kV. The conductor 1 to be studied is stretched along the axis of the outer cylinder 2 and connected to the alternating voltage $u(t)$. The current i in the earth lead of the insulated outer cylinder is measured. It may be assumed that this current approximately corresponds to that flowing from the high-voltage conductor. For exact measurements the corona-cage should be provided with a guard-ring arrangement.

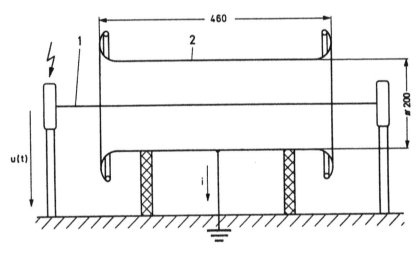

Fig.3.36 Corona-cage
1 Inner conductor, 2 Outer cylinder

The current i comprises of the displacement current and the corona current, with the capacitance thereby being assumed to be constant [*Sirotinski* 1955]:

$$i = C\frac{\mathrm{d}u}{\mathrm{d}t} + i_k$$

The corona current i_k increases rapidly with the instantaneous value of the voltage, once the onset voltage U_e is exceeded. It results from the migration of the ions formed by the discharge in the previous or in the same half-period. Fig. 3.37 shows the current characteristic to be expected for this considerably simplified treatment. The corona current i_k is real current and corresponds to the corona losses. These are caused by the power required to maintain collision ionisation as well as by the conductor current, represented by the movement of charge carriers. Corona losses in overhead lines are strongly dependent upon weather conditions and can deviate from the annual mean value up to one order of magnitude above or below.

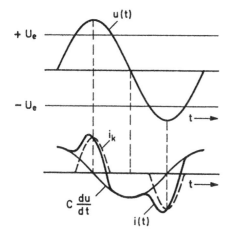

Fig.3.37 Voltage and current curves for the corona-cage

The charge carriers emerging from the collision ionisation region, by attachment to neutral gas molecules, form large ions which are accelerated away from the corona electrode; an "electric wind" is produced. This phenomenon has acquired great practical significance in the electrostatic purification of gases.

c) Partial Discharge Measurement in High-Voltage Insulation Systems
Partial discharges on or in a test object have become an important diagnosing means of high-voltage technology since they can be an indication of manufacturing defects in electrical equipment or the cause of ageing of an insulation. Details for the conduct of PD measurements in connection with insulation tests at alternating voltages are given in VDE 0434 and IEC Publ.270. For radio interference tests other aspects apply.

The most important PD measurements on high-voltage equipment aim to determine the onset voltage U_e and the extinction voltage U_a. In practical arrangements however, the onset and extinction of partial discharges are usually not very distinct phenomena. These measurements therefore require an agreement on the sensitivity of the methods used.

If a large number of PD sites are present in an insulation system, a noticeable increase of the losses in the dielectric occurs when the onset voltage range is exceeded. The magnitude of this increase is a measure of the intensity of the partial discharges, so long as the basic dielectric losses are low or remain constant. The Schering bridge is therefore also used for the measurements of corona losses in overhead lines or for the measurement of ionisation losses in cables, when these contain numerous distributed defects as a consequence of the manufacturing process (non-draining compound-filled cables).

To record and assess PD in technical insulation systems with isolated defects, more sensitive measuring methods should be applied. For this purpose instruments are used which amplify the high frequency electrical disturbances initiated by the partial discharges, and evaluate these in various ways. The measuring instrument is coupled as a rule by an ohmic resistance R, connected either to the earth lead of the test object as in Fig. 1.75 or to that of a coupling capacitor.

In a real test object, the voltage at R, as a consequence of the partial discharge, consists of an irregular train of pulses of very different amplitudes; their duration is dependent upon the characteristics of the circuit and can be some tens of ns. The objective of the PD measuring technique is to register this statistical quantity and evaluate it in view of the evidence desired. Various evaluation methods have been recommended for this purpose, among which the small-band or the wide- band charge measurement has emerged acceptable in practice. By calibrating with pulse generators, one aims to estimate the effect of the characteristics of the total setup upon the measured result.

One method, often adopted in testing bays predominantly while testing transformers, makes use of selective interference voltage measuring instruments developed as per VDE 0876 for measurement of radio interference voltages, for evaluating the measured quantity at R. These instruments are constructed as shown in the block circuit diagram of Fig. 3.38; evaluation follows taking into account the physiological impact of the disturbance on the human ear.

d) Gliding Discharges

One may always expect gliding discharges when high tangential field strengths appear at interfaces. For some insulating assemblies in high-voltage

to *TR*

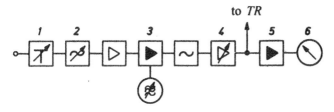

Fig.3.38 Block circuit diagram of a selective interference voltage measuring instrument
1 Input with variable damping element (calibration divider)
2 Tunable input circuit (filter)
3 Amplifier, oscillator, mixing stage
4 Intermediate frequency amplifier with variable amplification
5 Evaluating element
6 Indicating arrangement

technology, a flashover can be induced for this reason. Two typical examples of this are shown in Fig. 3.39. The basic shape of the equipotential lines can be illustrated by the partial capacitances pertaining to a virtual intermediate electrode A. Since the surface capacitance C_0 is much greater than C_1, almost the entire voltage appears on C_1.

When the onset voltage is exceeded, partial discharges occur which develop with increasing voltage from corona to brush discharges along the surface. The intensity of these gliding discharges and their onset voltage depend upon the magnitude of the surface capacitance C_0. The larger it is, the larger too, for time-varying voltages, is the discharge current which

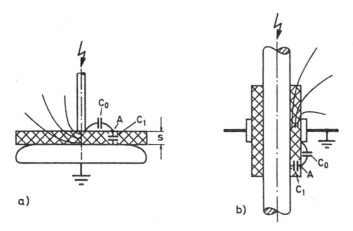

Fig.3.39 Gliding discharge arrangements
a) rod-plate, b) bushing

flows from the tip of the brush discharge through the insulator as a displacement current. This leads to extension of the high-voltage potential on the surface, without an appreciable reduction occurring in the field strength at the tip of the discharge. Further growth of the discharge is thus favoured.

Under direct voltages, gliding discharges occur, if at all, as very weak discharges owing to the absence of displacement currents. The decisive role is played here by the surface conductivity.

Under impulse voltages, the rapid voltage variations lead to particularly large displacement currents, which is why the gliding discharges in this case have a very high energy. From the shape and range of the gliding discharges, it is possible to deduce the polarity and the amplitude of an impulse voltage ; this fact is made use of for measuring purposes in Klydonographs. Here, in an electrode arrangement similar to Fig. 3.39a with a point high-voltage electrode, the upper surface of the insulating plate is coated with an active photochemical or dust-like layer. Lichtenberg figures, two examples of which are reproduced in Fig. 3.40, are obtained in this way. These show clearly the distinct polarity dependence of the gliding discharge mechanism [*Marx* 1952 ; *Nasser* 1971].

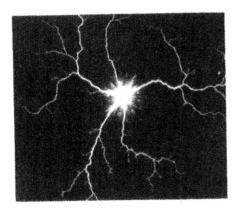 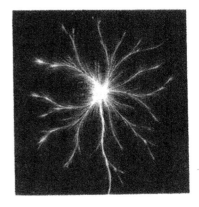

Fig.3.40 Lichtenberg figures (after *Marx* 1952)
a) positive point, b) negative point

The determination of the onset voltage U_e for the different discharge phases in a gliding discharge arrangement at alternating voltages, is of particular significance to the design of an insulating system. As shown by *M.Toepler* in 1921, U_e decreases with increasing magnitude of the surface capacitance. For the plane configuration with sharp-edged high-voltage electrode, as in Fig. 3.39a, the following empirical relationship is valid, with U_e in kV and s in cm [*Kappeler* 1949; *Böning* 1955; *Kind, Kärner* 1982]:

$$U_e = K\left(\frac{s}{\varepsilon_r}\right)^{0.45}$$

The values of K depend on material and are different for each discharge phase. They are, approximately:

Corona onset: $K = 8$ for metal edge in air
                    $K = 12$ for graphite edge in air
                    $K = 30$ for metal or graphite edge in oil
Brush discharge onset: $K = 80$ for metal or graphite edge in air or oil

Overstepping the brush discharge onset voltage often leads to permanent damage of the insulating surface within a very short time.

3.6.2 Experiment

The following circuit elements are used repeatedly in this experiment:

T   Testing transformer 220V/100 kV, 5 kVA
SM   Peak voltmeter (see 3.1)
CM   Measuring capacitor 100 pF, 100 kV
TR   Transient recorder
STM Interference voltage measuring device STTM 3840 a
     (manufactured by Siemens),
     measuring frequency 30 kHz to 3 MHz (set to 1.9 MHz),
     bandwidth 9 kHz
AV   Coupling four-pole STAV 3856 (60 Ω)

All measurements are done with alternating voltage as per the circuit in Fig. 3.41, however with different specimen.

a) Partial Discharges at a Needle Electrode in Air
A needle-plate gap, as in Fig. 3.34, with spacing $s = 100$ mm is incorporated as test object. The high-voltage electrode consists of a rod with a conical tip, into which a sewing needle has been inserted. The interference voltage measuring device should be connected to the four-pole AV in the earth lead of the test object. The various interference voltage pulses could be taken from the intermediate frequency amplifier of the device constructed according to Fig.3.38. By time dilation of the signal, a convenient

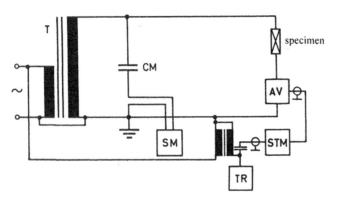

Fig.3.41 Test circuit for partial discharge measurements

oscillographic indication of the onset of pulses is possible. Corresponding to Fig. 3.41, the pulses are capacitively superimposed on an alternating voltage in phase with the test voltage, so that their phase relation with respect to the test voltage can be shown on the TR. A tolerably good recording of the pulse shape itself calls for measuring equipment with bandwidths of at least 100 MHz.

The discharge patterns at the needle for each voltage range are observed by varying the test voltage, and compared with the schematic representation of Fig. 3.35.

b) Measurements in the Corona-Cage

The corona-cage as in Fig. 3.36 should now be connected as the test object. A bare copper wire of diameter $d = 0.4$ mm is inserted as the inner electrode. In a first series of measurements the PD interference voltage U_{PD} ($= U_{TE}$ in German text) is measured as a function of the test voltage. At the same time the phenomena of incomplete breakdown should be observed up to the onset of complete breakdown.

In a second series of measurement, the coupling four-pole is replaced by a screened measuring resistor, to which a capacitor and a surge diverter are connected in parallel for overvoltage protection. The time constant RC should be about 100 μs.

At increasing test voltage, the time-dependent curve of the cage current i should be observed for up to about 80% of the breakdown voltage, and recorded at a voltage U which produces a particularly distinct curve. Fig. 3.42 shows an oscillogram of the cage current at $U = 40$ kV. The curve confirms the ideas described in 3.6.1b.

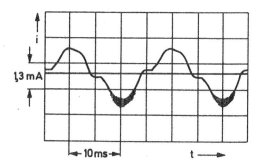

Fig. 3.42 Oscillogram of the cage current
$U = 40$ kV, diameter of the inner conductor $d = 0.4$mm

c) Partial Discharge Measurements on a High-Voltage Apparatus
A 20 kV current transformer should be connected as test object ; its high-voltage terminals should be fitted with screening electrodes if necessary, to avoid external partial discharges. The earthing connection of the test object is again made via the coupling four-pole ;the interference voltage measuring device is connected.

The PD interference voltage U_{PD} should be measured for up to 90% of the test voltage stated on the rating plate of the test object. The voltage should then be reduced at about the same rate and in doing so U_{PD} determined again. U_e and U_a should be measured. The curves shown in Fig. 3.43 were obtained for this experiment.

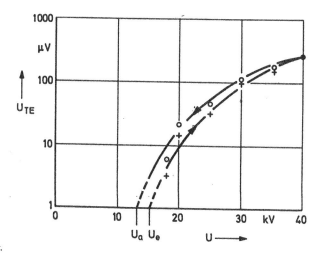

Fig. 3.43 Curve of the interference voltage of a 20 kV current transformer

d) Measurement of the Onset Voltages of Gliding Discharges
The test object should be arranged according to Fig. 3.39a with glass plates
in air as the dielectric. The relationship $U_e = f(s)$ should be measured for
various plate thicknesses s = 2, 3, 4, 5, 6, 8 and 10mm. The onset of the
gliding discharges in Fig. 3.44 should be determined with the STM and
that of the brush discharges visually.

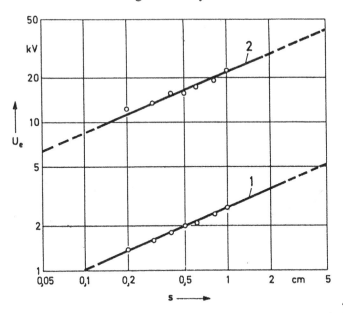

Fig. 3.44 Onset voltage of a gliding discharge arrangement as in Fig. 3.39a
1 Corona onset, 2 Brush discharge onset

By logarithmic graduation of the coordinates, the measured points can
be represented quite well by straight lines. This corresponds to the
relationship given in 3.6.1d :

$$U_e \sim s^{const}.$$

With $\varepsilon_r \approx 10$, for the straight line 1, one obtains $K \approx 8$ and for the straight
line 2, $K \approx 70$. Deviations to higher values for the corona onset voltage
can occur for insulating materials with high surface resistance, such as e.g.
glass, and this may be explained by the formation of surface charges.

3.6.3   Evaluation

Using the measurements from 3.6.2b, the breakdown strength E_d of the
wire used as the inner conductor of the cage should be calculated.

From the time variance of the current *i* recorded as in 3.6.2b, the approximate separation of the two components according to 3.6.1b should be attempted.

From the results obtained in 3.6.2c, the relation U_{PD} = $f(U)$ should be plotted in a diagram for increasing and decreasing test voltages of the current transformer. The relation U_e = $f(s)$ measured in 3.6.2d for the onset of corona and brush discharges should be plotted in double-logarithmic representation.

Literature : *Gänger* 1953; *Sirotinski* 1955 ; *Roth* 1959 ;*Schwab* 1969; *Nasser* 1971 ; *Bartnikas, McMahon* 1979; *Kind, Kärner* 1982

3.7 Experiment " Breakdown of Gases"

The analysis of the breakdown of gases is also important for understanding the breakdown mechanisms in liquid and solid insulating materials. Gas discharges always occur after the onset of breakdown in any type of dielectric. Gases have a wide range of application as insulating media, especially atmospheric air. The topics covered by this experiment fall under the following headings:

- Townsend mechanism,
- Streamer mechanism,
- Insulating gases.

It is assumed that the reader has some basic knowledge of

- the mechanisms of electrical breakdown of gases, as well as
- acquaintance with section 1.1 "Generation and measurement of alternating voltages".

3.7.1 Fundamentals

a) Townsend Mechanism
The breakdown of gases at low pressures and small spacings can be described by the Townsend mechanism. Thereby, electrons of external origin accelerated by the field can form new charge carriers by collision ionisation, provided their kinetic energy exceeds the ionisation potential of the gas molecules concerned. An electron avalanche is built up which travels from

the cathode to the anode. If, as a consequence of the avalanche, a sufficient number of new ions are formed near the cathode, complete breakdown finally takes place.

It can be shown that for this kind of discharge formation the static breakdown voltage U_d of a homogeneous field at constant temperature depends only upon the product of pressure p and spacing s. The ionisation coefficient of the electrons α and its dependence upon the field strength E can be described by the formula :

$$\frac{\alpha}{p} = Ae^{-B\frac{p}{E}}$$

where A and B are empirical constants. For the Townsend mechanism in a homogeneous field, the following breakdown condition is valid :

$$\alpha s = k = const.$$

When this equation is satisfied, $E = E_d = U_d / s$. Substituting and solving for U_d, one obtains the Paschen law:

$$U_d = B \frac{ps}{\ln\left(\frac{A}{k} ps\right)} = U_d(ps)$$

Whether or not the conditions of this law are satisfied can be taken as evidence for or against a discharge occurring by the Townsend mechanism.

b) *Streamer Mechanism*
At higher pressures and larger spacings discharge in gases takes place by the streamer mechanism according to *Raether, Loeb* and *Meek*. It is characterised by the fact that photon emission at the tip of an electron avalanche induces and initiates the growth of a streamer at a very fast, abrupt rate, compared to the growth of the primary avalanche.

The onset of photo-ionisation, very effective for the growth of the discharge, should be expected when the multiplication factor of the avalanche, $e^{\alpha x}$, has reached a critical value of about $e^{20} \approx 5.10^8$.

The transition of a discharge from Townsend growth to streamer growth can, for a given spacing, be promoted by several parameters.

The larger the product ps, the smaller is the probability that an individual avalanche can traverse the discharge space before critical multiplication is reached. For overvoltages up to about 5% above the static breakdown value of U_d, a discharge in air by the Townsend mechanism may be expected only for values of

$ps \leq 1$ MPa mm.

At higher values, breakdown occurs by the streamer mechanism.

For steep impulse voltages, high field strengths can appear locally which lie well above the static value of E_d, depending upon the impulse voltage-time curve of the arrangement. α increases strongly with E and consequently critical multiplication will be reached even in a short avalanche length.

The ionisation probability of photon radiation is approximately proportional to the density of the gas. Therefore the greater the product of molecular weight M and pressure p, the sooner critical multiplication of the avalanche and with that its change-over to streamer growth takes place.

High field strengths already prevail in strongly inhomogeneous fields near electrodes with strong curvature before the ignition of a self-sustained discharge. Thus it can be shown that for spherical and cylindrical electrodes, E_d increases rapidly with decreasing radius of curvature r. It follows that an avalanche, once started, easily reaches critical multiplication.

c) Types of Gas Discharges

The resistance of a gas-filled gap collapses to low values once the voltage for complete breakdown is reached. The type of gas discharge which then occurs and its duration depend upon the yield of the energising current source. When currents of the order of 1 A or more flow in the discharge path, one may expect arc discharges. In this case a well-conducting plasma column develops as a result of thermo-ionisation, the arc voltage of which decreases with increasing current.

If the current flowing after breakdown lies in the mA-range, one may expect glow discharges, particularly at low gas pressures (e.g. 10 kPa). For this type of discharge the charge carriers are formed by secondary emission at the cathode. A general statement concerning the current dependence of the arc voltage cannot be made.

The discontinuous transition to a discharge with higher current is referred to as spark discharge. In breakdown processes this is usually the transition to the arc discharge, which only lasts for a short while during voltage tests however. On the other hand, in power supply networks the extinction of an once established arc is usually only effected after switch-off.

d) Gases of High Breakdown Strength

Dry air or nitrogen are cheap insulating materials of high electrical strength, particularly at high pressures, which therefore find extensive technical applications. One may mention metal clad switch gear, compressed-gas capacitors or physics apparatus as examples. In all these cases, however, the mechanical stress to which the large containers are subjected calls for considerable constructional measures.

For homogeneous or only slightly inhomogeneous electrode configurations in air or nitrogen in the usual range of gap spacings of the order of centimetres, an increase in pressure beyond about 1 MPa results in progressive deviation from the Paschen law. The breakdown voltage U_d no longer increases in proportion with p, as shown in Fig. 3.45a. The reason for this probably rests with the associated ideas mentioned under 3.7.1b. For extremely inhomogeneous configurations a pressure increase could even lead to reduction of U_d. In this case promotion of the discharge growth by photo-emission predominates over the obstruction of collision ionisation owing to the increased pressure. Fig. 3.45b shows a schematic representation of the possible curve.

The excellent properties of sulphurhexafluoride (SF_6) for insulation and

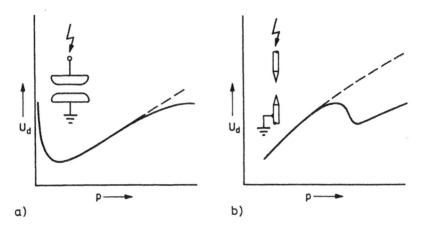

Fig. 3.45 Breakdown voltage of a gas as a function of pressure
Dotted line indicates behaviour according to the Paschen law
a) homogeneous field , b) inhomogeneous field

for arc-quenching have been known for a long time. Nevertheless, widespread application of this highly electro-negative gas has been in progress only since about 1960. It is used for the insulation of high-voltage switchgear, high-power cables, transformers and large-size physics equipment, as well as arc-quenching in power circuit breakers.

SF_6 has a molecular weight of 146 and is composed of 22% by weight of sulphur and 78% of fluorine. It is built up in such a way that the sulphur

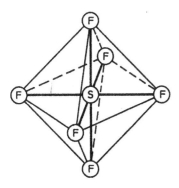

Fig. 3.46 Structure of an SF_6 molecule

atom is at the centre of a regular octahedron, with fluorine atoms at each of the six corners (Fig. 3.46). The ionisation energy of the process important for breakdown, namely :

$$SF_6 \rightarrow SF_5^+ + F^-$$

is 19.3 eV.

Sulphurhexafluoride, with density 6.139 g/l at 20°C and atmospheric pressure, is one of the heaviest gases and is 5 times as heavy as air. It is colourless, odourless, non-toxic and chemically very inactive. Since SF_6 has no dipole moment ε_r is 1 and independent of frequency.

The electrical strength of SF_6 in a homogeneous electric field is 2 to 3 times that of air. Results of measurements show, however, that the discharge growth in SF_6 can also be described reasonably well using the concepts of classical gas breakdown theory. This is shown by the pressure dependence of the breakdown voltage. The transition from the Townsend mechanism to the less advantageous streamer mechanism is expected for a very much lower pressure in SF_6 than in air. This is also especially true for the reduction of U_d in a strongly inhomogeneous field, shown in Fig. 3.45b [*Hartig* 1966].

During arc discharges in SF_6 reactive and toxic by-products are formed, which have to be absorbed by suitable agents (e.g. Al_2O_3).

3.7.2 Experiment

The data of the most important circuit elements are :

    T   Testing transformer 220V / 200kV , 10kVA

> CM   Measuring capacitor 200kV, 100pF
> G   Rotary vacuum pump, model D 6[3]

a) Experimental Setup
The experiments are performed with the setup shown in Fig. 3.47. The
alternating test voltage obtained from the test transformer T should be
measured using the peak voltmeter SM (see section 3.1) via different
measuring capacitors CM. The vacuum necessary for the experiment is
generated by the rotary pump G and measured by a membrane vacuum
meter M.

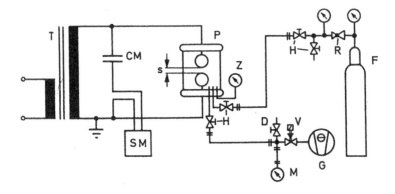

Fig. 3.47 Test setup for measuring the breakdown voltage at pressures of 0.1 kPa to
600 kPa

The regulating valve D is used for exact regulation of the desired pressure.
For measurements in the high pressure range (**Attention** :Limitation owing
to the mechanical strength of the pressure vessel !), a gas cylinder F with
a reducer valve R should be connected. (**Attention**: The gas cylinder should
be securely fixed to prevent toppling!). The high pressure is measured using
the indicating manometer Z mounted on the pressure vessel. Before
beginning the high-pressure experiments, one should make sure that the
membrane vacuum meter M is disconnected, to avoid damage. The stop-
cocks H allow connection of the required pipe-lines ; the magnetic valve V
closes automatically when the pump is switched off, so that unintentional
aeration of the container is prevented. The testing arrangement P is set up
in a pressure vessel as shown in Fig. 3.48. The insulating tube is of perspex
and thus permits visual observation of the discharge phenomena. The
electrodes can be exchanged by means of the removable insets; as an

---

[3] Manufacturer: M/s Leybold, Köln

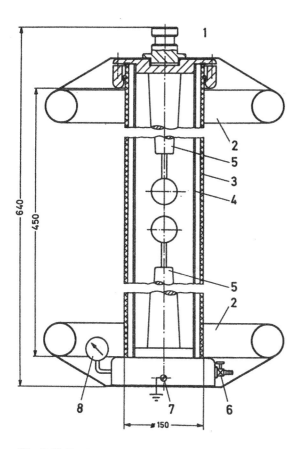

Fig. 3.48 Testing vessel for measurement of the breakdown voltage at pressures of 0.1 kPa to 6oo kPa

1	Vessel lid (high-voltage terminal)
2	Grading electrodes
3	Perspex cylinder
4	Hardboard cylinder
5	Electrode support
6	Stopcocks
7	Earth terminal
8	Pressure gauge

example, the figure shows an arrangement of two spheres of diameter $D = 50$ mm and spacing $s = 20$ mm, the most commonly used for the experiments. The pressure vessel is suitable for the proposed pressure range of about 0.1 kPa to 600 kPa and withstands a test pressure of about 1 MPa. The ring-shaped grading electrodes shown are necessary for

increasing the external flashover voltage. In this way measurements up to
200 kV a.c. could be carried out with this testing vessel.

b) Validity of the Paschen Law for an Electrode Configuration in Air
The electrode system to be investigated is a sphere-gap with $D = 50$ mm.
The a.c. breakdown voltage U_d in air should be measured for spacing $s =$
10 mm and 20 mm. The relation shown in Fig. 3.49 was obtained for the
described experiment. From this it follows that the conditions of the Paschen
law are well satisfied. Furthermore, diverse types of gas discharge occur
after breakdown in the investigated pressure range. Fig. 3.50 shows a glow
discharge at about p = 1 kPa and an arc discharge at normal pressure.

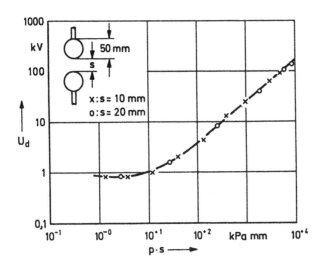

Fig. 3.49 Measured values of breakdown voltage between spheres in air

c) Breakdown Voltage of an Electrode Configuration in SF₆
With the aid of a second testing vessel as in Fig. 3.48, comparative
measurements of the breakdown voltage U_d of the sphere-gap should be
carried out in SF₆ at spacing $s = 20$ mm and for a pressure range of 100 to
250 kPa . The gas pressure is produced by an SF₆ compressed gas cylinder.

It is recommended that the measurements in SF₆ and in air be conducted
in separate testing vessels, because once a vessel is filled with SF₆ the
residual gas would continue to affect the results of later measurements in
air, despite long evacuation periods.

For measurements performed with the test system described, the values
indicated in the diagram of Fig. 3.51 were obtained. At the same pressure,
the strength of SF₆ is a factor 2 to 3 greater than that of air.

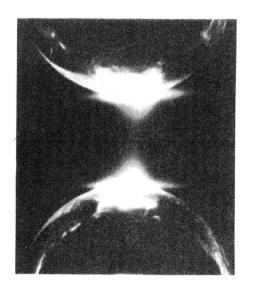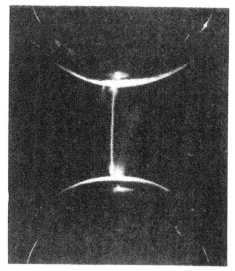

Fig. 3.50 Types of gas discharges in air at alternating voltages, (spacing s = 20mm)
a) glow discharge, pressure 1 kPa , exposure 5 s
b) arc discharge, pressure 100 kPa , exposure 40 ms

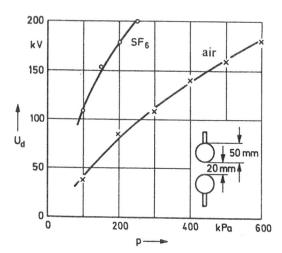

Fig.3.51 Pressure dependence of the breakdown voltage of a sphere-gap in air and in sulphurhexafluoride.

d) Pressure Dependence of the Breakdown Voltage in a Strongly Inhomogeneous Field

To demonstrate the breakdown performance of SF_6 as a function of pressure in a strongly inhomogeneous field, a point-plane electrode configuration is chosen. The diameter of the plate is $D = 50$ mm and the point is a $10°$ cone cut out of a 10 mm diameter rod. The gap spacing s should be set to 40 mm and measurements conducted in the pressure range of 100 to 600 kPa.

For the above experiments the relation shown in Fig.3.52 was obtained for the spacings $s = 20$, 30 and 40 mm. The falling tendency of the breakdown voltage at increasing pressures within a certain range, lies at appreciably lower values of pressure for heavy gases such as SF_6 than for lighter gases such as air. This effect may be accounted for by a change in the discharge mechanism, namely by the transition from the Townsend mechanism to the streamer mechanism [*Hartig* 1965].

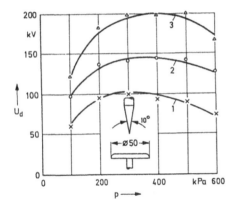

Fig. 3.52 Pressure dependence of the breakdown voltage of a point-plane gap in SF_6
1 $s = 20$ mm
2 $s = 30$ mm
3 $s = 40$ mm

3.7.3   Evaluation

The breakdown voltages as the function $U_d = f(ps)$ for the sphere-gap measured under 3.7.2b at spacing $s = 10$ mm and 20 mm at different pressures should be represented in a diagram on double-logarithmic paper.

The above values of the breakdown voltages U_d of the sphere-gap for $s = 20$ mm in air, together with the values measured in SF_6 under 3.7.2c, should be represented in a diagram as $U_d = f(p)$.

The breakdown voltages of the point-plane system in SF$_6$ measured under 3.7.2d should be shown as a function of pressure $U_d = f(p)$.

Literature : *Gänger* 1953 ; *Meek, Craggs* 1953 ; *Sirotinski* 1955; *Llewellyn-Jones* 1957; *Flegler* 1964 ; *Raether* 1964 ; *Kind, Kärner* 1982; *Kuffel, Zaengl* 1984; *Beyer et al.* 1986

3.8 Experiment "Impulse Voltage Measuring Technique"

The time-dependent character of an impulse voltage is often appreciably affected by the properties of the test object connected to the generator. This is particularly true for the case of breakdown phenomena with resultant intentional or unintentional chopping of the impulse voltage. Further, knowledge of the impulse voltage-time curves of practical high-voltage equipment is important for coordination of insulation in systems. Oscillographic measurements of rapidly varying voltages are therefore indispensable to the assessment of test results. The topics treated in this experiment fall under the following headings:

- Multiplier circuit after Marx.

- Impulse voltage divider,

- Impulse voltage-time curves.

Prerequisite for successful participation is a familiarity with sections:

- 1.3 Generation and measurement of high-impulse voltages and

- 3.3 Experiment "Impulse Voltages"

In this experiment, the *Marx* circuit for the generation of impulse voltages shown in Fig. 1.37 is used. The data of the single stage equivalent circuit, enabling calculation of the voltage form, can be determined according to section 1.3.2.

3.8.1   Fundamentals

a) Elements of an Impulse Voltage Measuring System
The block-diagram of a complete impulse voltage circuit is shown in Fig. 1.55. The high impulse voltage $u_1(t)$ to be measured must first be greatly reduced by a voltage divider. From a tap on this divider a measuring voltage proportional to the high-voltage signal is fed through a measuring cable,

either to a transient recorder TR or to an electronic peak voltage measuring instrument. The load capacitor of the impulse generator itself is often concurrently used as a capacitive voltage divider. For an impulse generator set up as in circuit a, the discharge resistor can also be employed as a resistive voltage divider. However, these arrangements are only suitable for the determination of the peak value of a full or tail-chopped lightning impulse voltage of the form 1.2/50. They are less suited for measurement of impulse voltages chopped on the front. A voltage divider which does not need to serve a double purpose, can be adapted better to the requirements of measurement.

The signal level at the input of the transient recorder is in general very low compared to that of the impulse voltage to be measured. Therefore, potential differences in the earthing system and electromagnetic interference voltages can influence the measured signal appreciably. Particularly during measurement of chopped impulse voltages, special earthing and shielding measures are essential, e.g. accommodating the transient recorder in a measuring cabin.

On its way from the test object terminals to the transient recorder, the signal to be measured is distorted, namely in general the more so the higher the frequency components it contains. For a measurement of fast-varying impulse voltages therefore, it is essential that the transmission behaviour of the measuring system be checked.

In high-voltage technology the step function response of the entire measuring system is adapted as a measure of the fidelity of reproduction. A characteristic parameter of the step response is the response time T. For known electrical properties of the divider, it can either be calculated or determined experimentally at low or high-voltages. In this experiment the method of determining T using a test gap of exactly known impulse voltage-time curve, as described under 1.3.14 is applied.

b) Impulse Voltage-Time Curves

When electrode systems are stressed by impulse voltages of a certain form and higher peak values than necessary for breakdown, then these are referred to as over-shooting impulse voltages. In these tests, the higher the peak value $\hat{U}$ is of the unchopped full impulse voltage, the shorter the time t_d becomes to breakdown of the test object. These correlations are described by the impulse voltage-time curve $U_d = f(t_d)$, which is typical for a given system and voltage form ; whereby U_d is the highest voltage value prior to breakdown and t_d is the time interval between the start of the impulse (point 0_1 in Fig. 1.35) and the start of the voltage collapse. For every impulse voltage-time curve one should specify the set impulse voltage form as well as the polarity on which the given characteristic is based.

According to 3.3, the breakdown of a gap occurs only when a voltage greater than the static onset voltage U_e persists at the gap for periods longer than some of the statistical time-lag t_s and the formative time-lag t_a. Since the front-time of a lightning impulse voltage of a given form is independent of the peak value $\hat{U}$, the voltage rise becomes steeper with increasing peak value. Hence for greater steepness the voltage can increase further beyond U_e during the breakdown time-lag t_v; the increase U_d for higher overshooting voltages is thus explained.

The statistical time lag and the formative time-lag are not however independent of the applied voltage. In systems with homogenous or only slightly inhomogenous fields (example: sphere-gap), t_s and t_a decrease rapidly with increasing overvoltage $\hat{U} / U_e$. In a system with strongly inhomogenous field (example: rod-gap), the formative time-lag determines the total time lag and decreases, even at high overvoltages, only slowly compared with the homogenous field case. Consequently the drop of the impulse voltage-time curve for short breakdown times is more pronounced for a system with an inhomogenous field than with one with a homogenous field.

The experimental determination of the impulse voltage-time curve of a given electrode configuration requires numerous individual measurements with various types of voltage forms. For this reason many authors have tried to determine the impulse voltage-time curve by calculation, using assumptions based on physical reasoning. Investigations have shown that such assumptions are valid only for a limited number of cases [*Wiesinger* 1966 ; *Hövelmann* 1966]. Even so, calculation of the impulse voltage-time curve under certain restricting conditions and for a particular range of breakdown times is still meaningful ; it offers the possibility of converting an impulse voltage-time curve measured for a similar configuration.

For calculation of the impulse voltage-time curve of electrode configurations with a homogenous or slightly inhomogenous field, the assumption that for a given gap the "formative area", i.e. the voltage-time area F above the static breakdown voltage U_e, remains constant even for different voltage forms [*Kind* 1958, *Kind, Kärner* 1982] has proved useful:

$$F = \int_{t_e}^{t_d} [u(t) - U_e] \mathrm{d}t = \text{const.}$$

The lower integration limit is fixed here by $u(t_e) = U_e$. These relations are illustrated in Fig. 3.53 for stressing by linearly rising impulse voltages with various gradients.

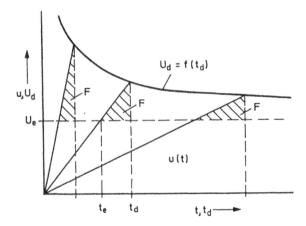

Fig. 3.53 Formative area and impulse voltage-time curve as per the "area rule"

If the formative area of a system is known by measurement with a particular voltage form, the breakdown voltage for any other voltage form can be calculated ; this is particularly easy for the case of linearly rising impulse voltages. Here, for rate of rise S, one has:

$$F = \frac{1}{2} \cdot \frac{(U_d - U_e)^2}{S}$$

$$U_d = U_e + \sqrt{2FS}$$

For the test gap mentioned in 1.3.14, the formative area $F \approx 2$ kV µs.

The statistical variations of the breakdown time-lag have so far been neglected. In reality, the impulse voltage-time curve is not obtained but a band of voltage-time curves, whose upper limit corresponds to a breakdown probability of 100%, referred to as statistical time lag curve. The lower limit is termed formative time lag curve and corresponds to a breakdown probability of 0%.

To guarantee effective insulation coordination for overshooting impulse voltages, the impulse voltage-time curve of an overvoltage protective device must lie below that of the equipment requiring protection, for all kinds of voltage gradients. This is generally ensured when surge diverters are used. However, if instead of the diverter a rod gap is used as a protective gap, the safety of the equipment is no longer guaranteed. The band of impulse voltage-time curves of a rod gap rises rapidly with the rate of rise of the voltage, whereas the voltage-time curve of an internal insulation, experimentally determinable only for simple models, can be flat even for very high rates of rise.

3.8.2 Experiment

a) Setup and Investigation of a Two-Stage Impulse Generator
Using the high-voltage construction elements, a two-stage impulse generator should be set up as in circuit a, to generate a positive 1.2/50 impulse voltage. The spatial arrangement recommended for the elements is shown in Fig. 3.54.

The elements required have already been mentioned under experiment 3.3. Apart from most of the elements in duplicate, on account of the two - stage construction, two additional charging resistors RV = 50 kΩ are required. The junction point between the RE is not connected with the rest

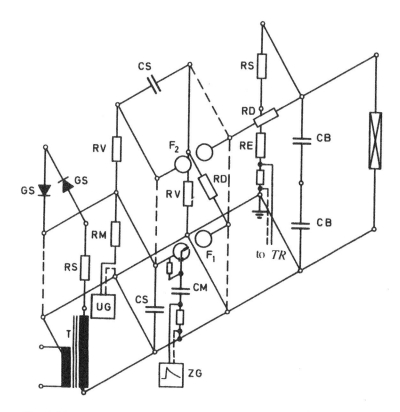

Fig. 3.54 Spatial arrangement of the elements of the two-stage impulse voltage generator

of the circuit ; this permits a simultaneous view of the discharge resistances as impulse voltage divider. As transient recorder, a TR with a bandwidth of about 50 MHz is suitable. If in case no special impulse voltage oscilloscope is available, it is advisable to make use of a measuring cabin (see 2.2).

The impulse generator is triggered by a pulse on F_1. The discharge resistance 2RE =19 kΩ, in this circuit parallel to the test object, is used as the high-voltage arm of a resistive voltage divider (divider I). The voltage at the low-voltage resistance arm of this divider is fed to the transient recorder TR by a coaxial, surge impedance $Z = 75$ Ω terminated measuring cable.

The faultless operation of the impulse generator, including the triggering of the transient recorder TR, should be checked over a large range of trigger gap spacings. Then two full impulse voltages with about 75 kV charging voltage per stage should be recorded with different time bases. In addition the peak value of this impulse voltage should be measured with a sphere-gap of $D = 100$ mm.

For this experiment the oscillogram shown in Fig. 3.55 was recorded, from which the time parameters can be taken as:

$$T_1 = 1.23 \ \mu s \qquad \text{and} \qquad T_2 = 45.6 \ \mu s$$

b) Comparison of the Fidelity of Two Impulse Voltage Dividers
A sphere-gap with $D = 100$ mm and $s = 30$ mm should be used as the test gap for comparing the fidelity of the two voltage dividers, as described in

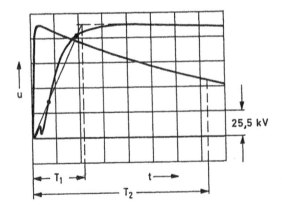

Fig. 3.55 Oscillogram of impulse voltage
Time-base : 0.57 and 5.7 μs / division
Charging voltage per stage : 75 kV
$T_1 = 1.23 \ \mu s$, $T_2 = 45.6 \ \mu s$.

1.3.14. In addition to the resistive voltage divider I, a damped capacitive voltage divider according to Fig. 2.12 (divider II), the construction of which is described in section 2.4, should also be investigated. When divider II is used, the connections are as in Fig. 1.54a.

For each divider, the time dependence of the voltage at the gap should be recorded in a common oscillogram, whilst the gap is stressed by three strongly overshooting impulse voltages. In doing so divider II should be connected in parallel to the test gap with a short lead of about 1 m length. Triggering of the TR is done with the aid of an antenna or directly by the trigger pulse for the impulse generator. Fig. 3.56 shows the " true" impulse voltage-time curve under standard conditions of the test gap used here. It was calculated by the method outlined in 1.3.14 for $F = 1.06$ kV μs.

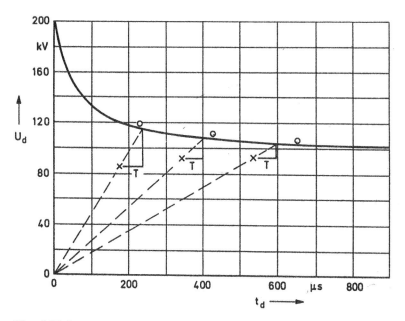

Fig. 3.56 Impulse voltage-time curve of the test gap (100 mm ϕ, $s=30$ mm) for positive wedge-shaped impulse voltages at standard conditions
x - measured points using divider I obtained at $d = 0.95$
o - measured points using divider II

Using the oscillograms, the measured breakdown voltage $(U_d)_{gem}$ of the test gap should be determined as a function of the rate of rise S of the voltage. This rise should be approximated as closely as possible by a straight line of slope S in the range between the measured breakdown voltage of the gap and its static breakdown voltage $U_e = 85.5$ kV. The point of intersection of this line with the base line shall be taken as the starting point of the idealised measured wedge-shaped impulse voltage. The time between this point and the voltage collapse is the measured breakdown time $(t_d)_{gem}$ of the idealised wedge-shaped impulse voltage.

The pair of values $(U_d)_{gem}$ and $(t_d)_{gem}$ contain amplitude and time errors, especially in the region of large rise. The rate of rise S shall however be assumed to have been measured accurately. The response times of the two voltage measuring systems should be determined by comparison of the measured values with the voltage-time curve of Fig. 3.56.

For this experiment the oscillograms shown in Fig. 3.57 were obtained for the two dividers. The static breakdown voltage of the test gap was determined in each case with a tail-chopped impulse voltage. The value pairs taken from the oscillograms of breakdown voltages and breakdown times for measurements conducted at $d = 0.95$, referred to standard conditions, are plotted in Fig. 3.56. The corresponding "true" points of the impulse voltage-time curve are the points of intersection obtained when a straight line of slope S passes through these measured points. The response time T and the voltage error ST can be read out directly.

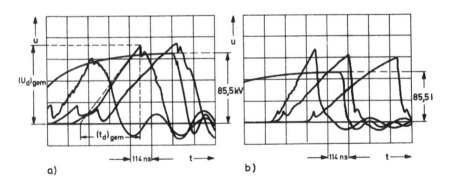

Fig. 3.57 Oscillograms of the voltage curves for stressing a sphere-gap of $s = 30$ mm in air, calibration with $U_{d-50} = 85.5$ kV
a) resistive divider I b) damped capacitive divider II

For the example shown here, the following mean values are obtained:

Divider I: T = 60 ns
Divider II: T ≈ 0 (evaluation uncertain).

c) Plotting Impulse Voltage-Time Curves
As in 3.8.2b, the test objects - a sphere-gap with $s = 45$ mm and a support insulator with a protective gap level corresponding to Series 10 N

(s = 86 mm) - should be investigated . Here too, the time dependence of three different overshooting impulse voltages should be recorded for each test object. The sphere-gap was chosen for these measurements because its impulse voltage-time curve is similar to that of the internal insulation of high-voltage equipment.

The breakdown voltage of each of the two investigated configurations should be determined from the oscillograms as a function of the breakdown time t_d. This shall be computed from the impulse start of the 1.2/50 standard impulse voltage. Divider II should be used for the measurements.

The oscillograms shown in Fig. 3.58 were recorded for this experiment. Their evaluation yields the impulse voltage-time curve of Fig. 3.59. From the point of intersection of the impulse voltage-time curves of the sphere-gap and the protective gap, the value of 0.12 kV/ns can be read out as a rough approximation for that rate of rise up to which protection is still provided.

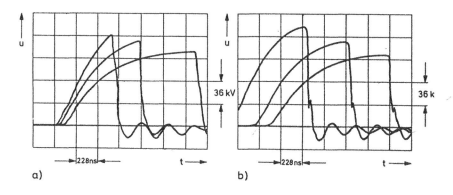

Fig. 3.58 Oscillograms of voltage forms for stressing of different test objects by overshooting impulse voltages
a) sphere-gap (s = 45 mm)   b) protective gap (s = 86 mm)

3.8.3   Evaluation

The data of the switching elements for the single stage equivalent circuit of the arrangement as in Fig. 3.54 and the utilisation factor η should be calculated.

T_1, T_2 should be determined according to 3.8.2a, $\hat{U}$ from the oscillograms; η should be determined and $\hat{U}$ compared with the result of the sphere-gap measurement.

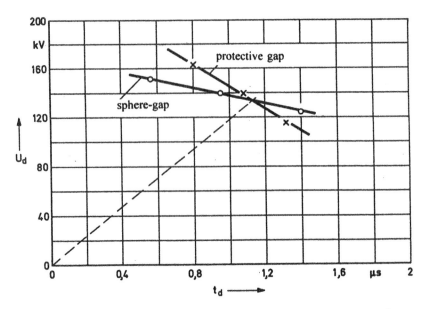

Fig. 3.59 Impulse voltage-time curves of a sphere-gap (D = 100 mm, s = 45 mm) and a protective gap (s = 86 mm)

The response time of the different dividers as in 3.8.2b should be compared. The impulse voltage-time curves should be plotted as per 3.8.2c.

Literature: *Strigel* 1955 ; *Schwab* 1969 ; *Kind, Kärner* 1982 ; *Kuffel , Zaengl* 1984

3.9 Experiment "Transformer Test"

The testing of technical products on the basis of certain specifications serves as a confirmation of agreed properties. Power transformers are important and costly elements in high-voltage networks; their reliable valuation by means of high-voltage tests is therefore of particular significance to the operational security of electrical supply systems.

The topics covered by this experiment fall under the following headings:

- Specifications for high-voltage tests,

- Insulation coordination,

- Breakdown test of insulating oil,

- Transformer test with alternating voltage,

- Transformer test with lightning impulse voltage.

It is assumed that the reader has :

- basic knowledge of the construction of 3-phase power transformers and familiarity with the sections

- 1.3 Generation and measurement of high impulse voltages

- 3.3 Experiment "Impulse Voltages"

3.9.1 Fundamentals

a) DIN / VDE Specifications and IEC Recommendations
Obligatory guidelines and regulations are necessary for the assessment of the quality of and the trade in electrotechnical products. At the national level in Germany these are provided and published by the Technical Committees of the Deutsche Elektrotechnische Kommission (German Electrotechnical Commission) in DIN and VDE (DKE). In order that they may not inhibit future development, the DIN/VDE specifications must be regularly revised and amended, to meet the corresponding status of technology. In addition to important safety regulations, they also contain instructions for conducting tests. In this way recognised rules of electrical technology resulted, which, in the event of damages, are also of legal consequence.

The need to publish similar specifications at the international level followed from the increasing expansion of trade beyond national borders. Due to the many differences prevalent amongst the nations on account of historical development, climatic variations and unit systems for example, international agreements can only take the form of overall recommendations. They are worked out by the Technical Committees of the IEC. The harmonisation of national specifications is of great significance to the economic cooperation between different countries.

VDE 0532 "Transformers and Choke coils", besides definitions of terms, contains specifications for the construction and testing of transformers. The high - voltage tests mentioned in these and other equipment regulations are in accordance with the rules concerning the magnitudes of test voltages (VDE 0111) and the generation and measurement of test voltages (VDE 0432).

b) Insulation Coordination
In the field of high-voltage technology, the " Specifications for Insulation Coordination" (IEC Publ. 71, VDE 0111) assume a special position, since

the rated withstand voltages and thereby the test voltages are specified there in a uniform way. Here, for identification of the insulation of an equipment, highest permissible operating voltages U_m have been specified; in 3-phase systems, it is the r.m.s. value of the maximum line-to-line voltage for which the equipment is designed.

In the case of external overvoltages, the definition of an insulation exempt from every risk is usually impossible, for economic reasons. The test voltage for lightning impulse voltages is therefore chosen so that no breakdowns can occur during operation either within the equipment or across open contacts. For insulation coordination it is essential that the strength of the internal insulation (upper impulse level) lies above the breakdown or flashover voltage of air gaps (lower impulse level). Further, the magnitude of overvoltages occurring must be limited by the use of overvoltage protective equipment (protective level). For lightning impulse voltages the voltage form is defined by 1.2/50.

The test using switching impulse voltages to verify the insulation strength against internal overvoltages has special significance for large spacings in air and a strongly inhomogenous field. Air spacings of insulation systems for operating voltages of over 245 kV should therefore be subjected to a corresponding type test. Switching impulse tests could also be effective as routine tests in lieu of a test with excessively high alternating voltages.

A few test voltages and protective levels for equipment in 3-phase systems with $U_m < 300$ kV are given in Table 3.1 as examples.

Table 3.1 Test voltages and protective levels for equipment in 3-phase systems (extract)

U_m	Alternating voltage	Lightning impulse voltage	Protective level
kV	kV	kV	kV
12	28	75	40
24	50	125	80
36	70	170	120
72.5	140	325	205
123	230	550	350
245	460	1050	750

c) *Testing of Insulating Oils*

Power transformers for high voltages contain large quantities of insulating oil for insulation and cooling. Good dielectric properties of the insulating oil are therefore an important prerequisite for perfect insulation of these transformers. Since the breakdown strength of an insulating oil depends appreciably upon its composition, preparation and ageing conditions, its determination is an important part of the high-voltage testing of transformers.

In VDE 0370 valid for insulating oils, a minimum quality is prescribed for new or used oils under exactly specified testing procedures. The complete testing programme covers, among others, the following properties : purity, density, viscosity, breakdown voltage, dielectric dissipation factor and specific volume resistivity.

The breakdown voltage should be measured using a standard testing vessel and alternating voltages of supply frequency. The spherical caps with spacing $s = 2.5$ mm shown in Fig.3.60 should be chosen as electrodes. The test voltage should be increased from zero at a rate of about 2 kV/s upto breakdown. Six breakdown experiments should be conducted for each oil sample. The mean value of the breakdown voltage determined from the 2nd to the 6th measurement may not be less than certain minimum values. These values are 60 kV for new oils in transformers and instrument transformers and upto 30 kV for switchgear ; lower values are permissible for equipment in service.

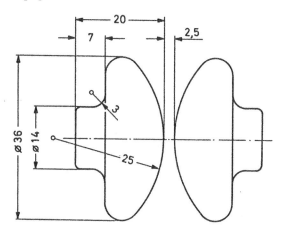

Fig. 3.60 Electrodes for the measurement of the breakdown voltage of insulating oils according to VDE 0370

d) Testing of 3-phase Transformers with Alternating Voltages
In high-voltage equipment with windings, one should distinguish between winding insulation tests and interturn insulation tests. Both tests are conducted as routine tests.

In the winding test at the test voltage U_p, the insulation between all the high-voltage windings, and the low-voltage windings connected to the core, is tested as shown in Fig. 3.61. Should the high-voltage windings be single-

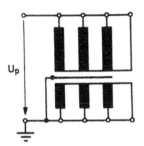

Fig. 3.61 Circuit for testing winding insulation

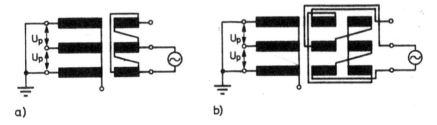

Fig. 3.62 Circuits for testing interturn insulation
a) vector group Yd5   b) vector group Yz5

pole insulated, the winding test on manufactured equipment can be carried out only at a voltage corresponding to the insulation of the earth-side terminal.

In the interturn test (induced voltage test), the mutual insulation of the individual turns is tested. In doing so the testing frequency may be increased, in case the current drawn is excessively large due to saturation of the iron core.

Two circuits are shown in Fig. 3.62 for the interturn test of 3-phase transformers with two different vector groups. The test should be performed by cyclic interchange of the phases. Excitation is thereby effected by connecting two terminals of the high-voltage or low-voltage winding to an adjustable alternating voltage.

e) Testing of Transformers with Lightning Impulse Voltages

For impulse voltage tests on transformers it is primarily the interturn test which is important, since an uneven voltage distribution along the winding may be anticipated (see also section 3.11). The particular difficulty of this test lies in the reliable identification of even small and only transient partial defects. On no account may a defect develop during the test which remains unidentified and could cause failure in service later on. As a rule, impulse voltage tests are conducted on transformers as type tests.

Fig. 3.63 shows a measuring circuit suggested by *R. Elsner* in 1949. The time variant form of the current i_c, which for fast processes is mainly capacitively transferred to the low-voltage winding US, is measured by the voltage drop it causes across the measuring resistor R_i. Partial breakdowns in the high-voltage winding OS modify the oscillations induced by the impulse and are further observed by the superposition of a higher frequency oscillation. Defects in the high-voltage winding, which occur depending on the amplitude of the impulse voltage, are identified by comparison of impulse voltage and impulse current curves obtained while testing with an impulse of sufficiently low amplitude(calibration impulse), low enough not to cause any defect, and on stressing with the full test voltage (test impulse).

A measuring circuit proposed by *J.H. Hagenguth* in 1944 is shown in Fig. 3.64. Here the magnetisation current i_o flowing from the stressed

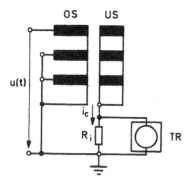

Fig. 3.63 Circuit for the lightning impulse voltage test according to *Elsner*
OS high-voltage winding ,   US low-voltage winding

winding to earth is measured. Fault identification again occurs by comparison of the oscillograms obtained during calibration and test impulses.

These tests are generally conducted with full impulse voltages. In special cases, a test with chopped impulse voltages can be additionally agreed upon with the customer. Because of the rapid voltage collapse this test represents an especially high stressing of the insulation.

A comparison of the curves of voltages and currents obtained during full and chopped impulses with varying times to chop is not possible, even after the collapse of the impulse voltage, for purposes of fault identification, due to wide variations in the forms of these curves.

If the impulse voltage $u(t)$ and the transmitted impulse current $i_c(t)$ are recorded with the digital recorder, digitised measured values are available for further processing with a computer whereby the transfer function of the transformer in frequency-domain, i.e. the quotient of the spectra of transferred impulse current and applied impulse voltage, can be calculated.

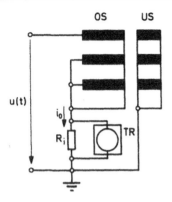

Fig. 3.64 Circuit for the lightning impulse voltage test according to *Hagenguth*
OS high-voltage winding   US low-voltage winding

For calculation of these spectra, Fast-Fourier-Transformation (FFT) can be applied [*Malewski, Poulin* 1985]. Fig. 3.65 shows the amplitude and the phase of the transfer function of a distribution transformer.

The transfer function of the transformer is independent of the time variant form of the test voltage and thus offers thereby the possibility to compare test impulses of different voltage forms and amplitudes with one another. If no fault is present in the transformer, the transfer functions of different impulses should be identical.

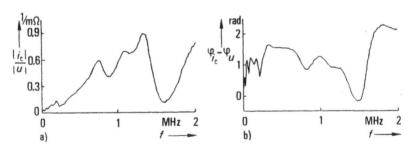

Fig. 3.65 Transfer function of an oil-immersed transformer (30kVA . 10kV / 400 V)

3.9.2   Experiment

a) Breakdown Test of an Insulating Oil According to VDE 0370
A circuit as shown in Fig. 3.66 should be set up. The following circuit elements will be used:

    T   Test transformer, rated transformation 220 V/100 kV
    CM   Measuring capacitor, 100 pF
    SM   Peak voltmeter (see 3.1)

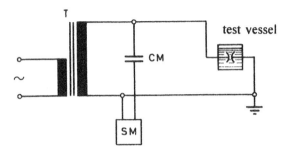

Fig. 3.66 Circuit for the breakdown test of insulating oils

An oil sample is taken from the transformer to be tested. The oil to be investigated should be poured slowly into the testing vessel, avoiding bubble formation (by allowing it to run along a glass rod), and then left to stand for about 10 min before the voltage is applied. The voltage should be switched off at the instant of breakdown. An interval of about 2 min should be maintained after each breakdown and the breakdown path between the electrodes flushed with new oil by carefully passing a stirring-rod through the gap.

b) AC Test of an Oil-Filled Transformer According to VDE 0532
In the circuit of Fig. 3.66, an oil-filled transformer of the voltage series 20 kV is connected as the test object.

As far as is practicable, the high-voltage winding of the transformer should be subjected to the a.c. test voltages. However, as specified for repeat tests on transformers in service beyond the guarantee period, only 75% of the test voltage values according to VDE 0111 may be applied here.

c) Impulse Voltage Test of an Oil-Filled Transformer According to VDE 0532
A single-stage impulse generator as in Fig. 3.13, but to generate negative lightning impulses of the form 1.2/50, should be set up as in VDE circuit a. The 3-phase transformer of 3.9.2b should be connected as the test object, with a rod gap of adjustable spacing s in parallel. Fault identification should be realised with the help of either of the circuits shown in Fig. 3.63 and Fig 3.64 (guiding value for $R_i = 75 \ \Omega$), using a 2-channel transient recorder (bandwidth $\geq$ 5 MHz). Satisfactory working of the setup, including the

oscillographic measuring setup, should be checked without the test object for d.c. charging voltages $U_0 = 70$ to 130 kV. The measurement of the peak value can be effected here via U_0 whereby a constant value is assumed for the utilisation factor η. The test object should then be connected. For verification of the upper and lower impulse levels, the following tests should be performed (though here as repeat tests at only 75% of the new values):

- 2 calibration impulses with full impulse voltages at 75% of the lower level values
- 2 test impulses with chopped impulses at the upper level values
- 2 control impulses with full impulse voltages at 100% of the lower level values.

In doing so, the time variant forms of the voltage and current should be recorded. Before connecting the test object, the spacing of the rod gap should be adjusted such that the impulse voltage at the upper level is chopped after about 2 ... 4 μs. The oscillograms reproduced in Fig. 3.67 were obtained for this kind of test. The conformity of the recordings in a) and c) shows that the transformer has passed the test.

If the time variant forms of the impulse voltage and impulse current are obtained with a digital recorder, the transfer function due to different impulses can be utilised to adjudge the condition of the insulation of the transformer.

3.9.3 Evaluation

The breakdown voltage of the oil investigated in 3.9.2a should be determined. Can this oil be used in new transformers?

Has the oil-filled transformer passed the a.c. test according to 3.9.2b?

What is the voltage taken by the neutral point of the high-voltage winding of a transformer during an interturn insulation test (Fig. 3.62) at 100% test voltage? What is the test frequency then required?

By comparing the oscillograms recorded under 3.9.2c one may determine whether the test was withstood.

Literature: *Wellauer* 1954; *Strigel* 1955; *Sirotinski* 1965; *Heller, Veverka* 1968; *Greenwood* 1971; *Malewski, Poulin* 1985

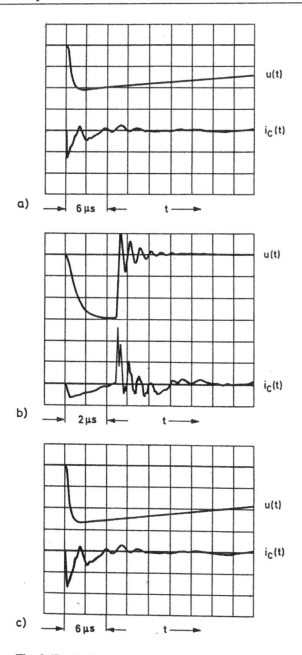

Fig. 3.67 Oscillograms of an impulse voltage test on an oil-filled transformer according to *Elsner*

a) calibration impulse

b) test with chopped impulse voltage

c) control with full impulse voltage

3.10 Experiment "Internal Overvoltages"

After switching operations as well as due to non-sustained earth faults and short-circuits in electrical systems, overvoltages caused by transient phenomena may occur; as a result the insulation of working equipment is in danger. Besides these transient overvoltages, permanent overvoltages too can occur in networks whose neutrals are not earthed. Noteworthy here are the power frequency oscillations in inductances with non-linear characteristics, which are designated as "Ferro-resonance". The topics covered in this experiment can be summarised under the following headings:

- Neutral shift,
- Earthing coefficient,
- Magnetising characteristic,
- Jump resonance,
- Subharmonic oscillations.

It is assumed that the reader has some basic knowledge of

- multi-phase systems and
- the properties of oscillatory circuits.

3.10.1 Fundamentals

a) Multi-Phase Networks with Non-Earthed Neutral
In 3-phase high-voltage transmission systems with non-earthed neutral point, additional overvoltages can occur as opposed to systems with earthed neutral point. They arise as a result of an earth fault of a conductor or are incited by switching operations as jump resonance and subharmonic oscillations. The reasons for these phenomena may be traced to inductances with non-linear behaviour. These could, for example, as main inductances of voltage transformers or power transformers, form oscillatory circuits with the earth capacitances of the network.

Overvoltage phenomena of this kind occur in a 2-phase network with isolated neutral in a manner analogous to a 3-phase system,. Since 2-phase network simulation is easier to realise and the processes during the generation of the overvoltage can be readily surveyed, a 2-phase system is used in the experiment discussed here. Fig.3.68 shows the equivalent circuit valid for basic studies of a 2-phase network, with the transformer *T*, neutral

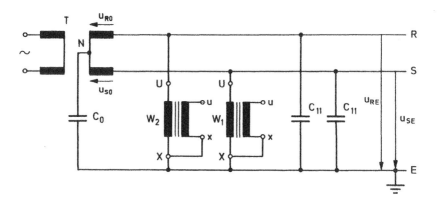

Fig. 3.68 Simplified representation of a 2-phase network with non-earthed neutral
T Transformer C_{11} Earth capacitances
C_0 Neutral capacitance W_1, W_2 Voltage transformers

capacitance C_0, earth capacitances C_{11} and the single-pole insulated voltage transformers W_1 and W_2 with terminal markings U-X and u-x for the primary and secondary windings respectively. The potential of earth is designated by E and that of the midpoint of the transformer by N.

For normal working conditions, the potential of N is determined by the value of the earth and neutral capacitances. Under symmetrical loading conditions and with equal earth capacitances C_{11}, no potential difference exists between mid-point N and earth E. The potential curves and phasor diagram of a 2-phase system for this working condition are shown in Fig.3.69. The phase voltages are displaced by 180° with respect to one another. The voltages u_{RO} and u_{SO} can be measured at the terminals u-x of the voltage transformers. When the symmetry of the system is disturbed the potential curves change. Fig.3.70 shows the voltages which would appear on earth fault of phase S. The transformer voltages u_{RO} and u_{SO} are fixed. As a result of the short-circuit the mid-point N is displaced with respect to

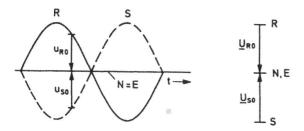

Fig.3.69 Potential curves and phasor diagram of a symmetrical 2-phase network

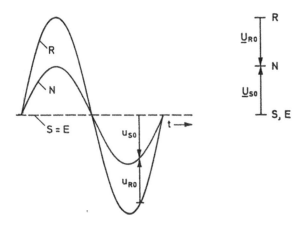

Fig.3.70 Potential curves and phasor diagram of a 2-phase network with an earth fault in phase S

earth E. At the terminals u-x of the voltage transformer of phase S the value measured is zero, and for phase R it is $u_{RO} - u_{SO}$.

In the case of an earth fault the insulation of the intact conductor as well as that of the midpoints of the transformers with respect to earth will be heavily stressed. To give an idea of this voltage increase, the earth fault factor δ has been introduced with the following definition (VDE 0111):

$$\delta = \frac{\text{Voltage of the healthy conductor against earth during short-circuit}}{\text{Phase voltage}}$$

For a multi-phase network with non-earthed neutral $\delta = \sqrt{3}$. A network with $\delta \leq 1.4$ is considered by definition to be solidly earthed.

The network shown with floating neutral in Fig. 3.68, with the capacitances C_0 and C_{11} and the current-dependent main inductances of the voltage transformers, represents a non-linear oscillatory circuit. Switching operations or fault conditions can induce the system to oscillate either by way of jump resonance or subharmonic oscillations, depending on the magnitudes of C_0 and C_{11}. An oscillation of the mid-point N with respect to the earth potential E also occurs at the same time.

b) Jump Resonance

For closer investigation of the natural oscillations the circuit of the 2-phase network shown in Fig. 3.68 is converted to equivalent circuit as in Fig. 3.71. For this purpose the network is assumed to be a linear active four-

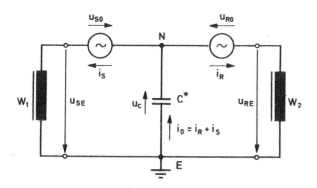

Fig.3.71 Equivalent circuit of the 2-phase network shown in·Fig.3.68.
C^* Neutral capacitance of the equivalent circuit
W_1 , W_2 Voltage transformers

pole, at the terminals of which the voltage transformers W_1 and W_2 are connected. According to the theory of the equivalent voltage source the capacitance of the 2-phase equivalent circuit is :

$$C^* = C_0 + 2C_{11}$$

The analogue treatment of the 3-phase network results in an effective equivalent capacitance between N and E :

$$C^* = C_0 + 3C_{11}.$$

The voltages $\underline{U}_S$ and $\underline{U}_R$ remain unchanged in the equivalent circuit, having the same magnitude and opposite phase. For given elements, the currents and voltages for the stationary condition can be approximately calculated [*Philippow* 1963].

With the currents and voltages shown in the equivalent circuit, the following loop equations are valid :

$$u_{RE} + \frac{1}{C^*}\int(i_R + i_S)\,\mathrm{d}t = u_{R0}$$

$$u_{SE} + \frac{1}{C^*}\int(i_R + i_S)\,\mathrm{d}t = u_{S0}$$

The network with two non-linear elements shown in Fig.3.71 can execute natural oscillations in various ways. The solutions of the system of non-linear differential equations possess great diversity, even when only rough approximations can be made for the characteristics of the two inductances

[*Knudsen* 1953 ; *Philippow* 1963 ; *Hayashi* 1964]. Hence only three simple limiting cases of the currents i_R and i_S, for which clear-cut solutions can be given shall be considered here as follows:

$i_R = - i_S$:   Phase opposition of currents is to be expected in a fully symmetrical circuit ($u_{R0} = - u_{S0}$). The effective capacitance between N and E always remains current free ($i_0 = 0$). Natural oscillations do not occur. This is the symmetrical operating condition.

$i_R = i_S$:   For in-phase currents, $i_0 = 2i_R$ and the first loop equation

takes the form $u_{RE} + \dfrac{1}{C} \int i_R dt = u_{R0}$ where $C = \dfrac{1}{2} C^0$

$i_R \gg i_S$:   For strong asymmetry of the two loops, $i_0 \approx i_R$ and the first loop equation takes the same form as for $i_R = i_S$, but with $C = C^*$.

In fact, the condition for complete symmetry is only insufficiently fulfilled solely because of the different characteristics of the voltage transformer inductances; the more the characteristics of the magnetic cores diverge, the greater the current i0. In this way, as a result of voltage increase or switching operations, the circuit can be induced to carry out neutral oscillations. The loop equation for the two limiting cases mentioned last is satisfied by the series oscillatory circuit shown in Fig.3.72, which is therefore suited for simple investigations of the oscillatory behaviour of networks with non-earthed neutral [*Knudsen* 1953 ; *Rüdenberg* 1953 ; *Philippow* 1963 ; *Hayashi* 1964 ; *Sirotinski* 1966].

In the following only the fundamental oscillations of the current and the voltage shall be considered. Here the phasor $\underline{U}$ represents the system voltage and $\underline{U}_L$ is the voltage across the inductance L. The magnitudes of the phasors are r.m.s. values; this will not be specifically pointed out each time.

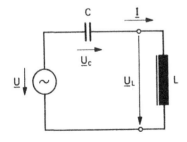

Fig.3.72 Undamped series oscillatory circuit with non-linear inductance

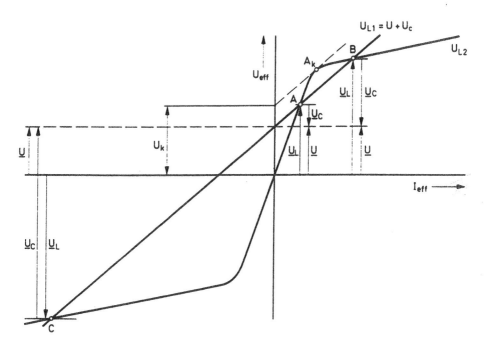

Fig.3.73 Graphical determination of the working points for the oscillatory circuit of Fig.3.72.

The behaviour of the single-phase oscillatory circuit will be explained with the help of the current-voltage characteristic shown in Fig.3.73. For this purpose let us imagine that the circuit of Fig.3.72 is interrupted at the terminals of L. Possible currents $\underline{I}$ are characterised by the condition that the voltage supplied by the circuit $\underline{U}_{L1} = \underline{U} - \underline{U}_c = \underline{U} - \underline{I}/j\omega C$ corresponds to the voltage $\underline{U}_{L2}$ across the inductance L, which is given by the r.m.s. characteristic.

For the working points A, B and C we have therefore:

$$\underline{U}_{L1} = \underline{U}_{L2} = \underline{U}_L$$

For clarity, the diagram also shows the voltages at the working points as phasors.

Investigation of the stability of the working points can be effected by the method of virtual displacement. For an imaginary increase of the current at the point A, $\underline{U}_L + \underline{U}_C > \underline{U}$. Since the driving voltage $\underline{U}$ is smaller than the sum of the voltages appearing across L and C as a result of the increased current, the current reverts to its original value. The same result is obtained for an imaginary reduction of the current; A is therefore a stable working point. At point B on the other hand, every displacement of the current results in a voltage difference, which strives to increase the deviation further. The condition at B is therefore unstable.

In the same way, it can be shown that C is another stable working point. Owing to the very much higher values of current and voltage compared with those at A, however, operation at point C would mean risk to circuit elements due to overvoltages and thermal overloading.

The jump resonance procedure can be triggered by switching operations, which cause a temporary increase of the supply voltage $U > U_k$ for example. Jump resonance can also be induced by switching off circuit components and by the associated reduction of the earth capacitance C_{11}; in this event the characteristic $\underline{U}_{L1}$ in Fig.3.73 would become steeper. The most important means of preventing jump resonance in networks is by introducing damping resistances in the secondary circuit of the voltage transformer, preferably connected to earth fault windings in series. In the single-phase equivalent circuit these resistances R act in parallel with L. This equivalent circuit and the corresponding phasor diagram are shown in Fig.3.74. For the magnitudes we have:

$$U^2 = \left(U_{L1} - \frac{I_L}{\omega C}\right)^2 + \left(\frac{U_{L1}}{R\omega C}\right)^2$$

$$= U_{L1}^2 \left(\frac{1+\alpha^2}{\alpha^2}\right) - 2U_{L1}\frac{I_L}{\omega C} + \left(\frac{I_L}{\omega C}\right)^2 \quad \text{with} \quad \alpha = R\omega C .$$

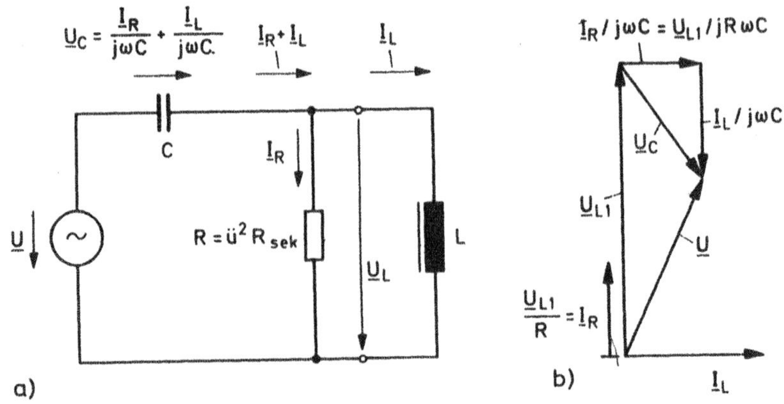

Fig. 3.74 Damped series oscillatory circuit with non-linear inductance
a) equivalent circuit, b) phasor diagram

The solution of this quadratic equation gives :

$$U_{L1} = \frac{\alpha^2}{1+\alpha^2} \cdot \frac{I_L}{\omega C} \pm \sqrt{\frac{\alpha^2 U^2}{1+\alpha^2} - \frac{\alpha^2}{(1+\alpha^2)^2} \cdot \left(\frac{I_L}{\omega C}\right)^2}$$

The square root expression represents an ellipse in the current-voltage plane with the semi axes $U\omega C\sqrt{1+\alpha^2}$ and $U\alpha/\sqrt{1+\alpha^2}$; the expression $\frac{\alpha^2}{1+\alpha^2}\frac{I_L}{\omega C}$ is a straight line.

The sum of the two functions is the sheared ellipse shown in Fig.3.75. The points of intersection A, B and C, as well as the stable working points A and C are again characterised by the relationship :

$$U_{L1} = U_{L2} = U_L.$$

For $R \to \infty$, the ellipse degenerates into two straight lines with slope $1/\omega C$ and points of intersection with the ordinate at $\pm U$. With decreasing R the ellipse becomes smaller, corresponding to higher damping, and rises less steeply, so that finally only one stable working point A remains and jump resonance can no longer occur.

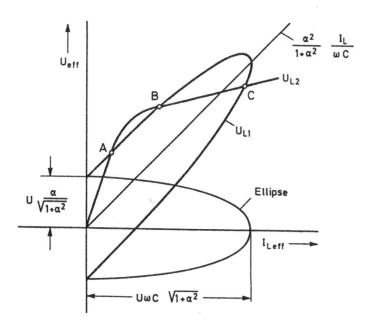

Fig 3.75 Graphical determination of the working points for the oscillatory circuit of Fig. 3.74

c) Subharmonic Oscillations

A further consequence of non-linear inductances in multi-phase systems with floating neutral is the occurrence of subharmonic oscillations. These are stationary oscillations whose frequency is an integral fraction of the supply frequency. In 50 Hz systems, frequencies of 25 and 16 2/3 Hz occur most commonly.

Although these phenomena are very complicated and difficult to calculate, an attempt shall be made here to give a clear explanation of the mechanisms which, to a limited extent, may even be quantitatively interpreted.

If the voltage transformer used in 3.10.2a is connected to a direst voltage source as in Fig. 3.74a, the curves of the voltage $u_L(t)$ and the current $i_L(t)$, recorded in the oscillogram of Fig. 3.76 are obtained. Since for saturation a constant voltage-time area is always required, the period increases with decaying amplitude.

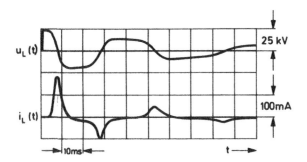

Fig. 3.76 Natural oscillations of the non-linear oscillatory circuit of Fig. 3.74

Such inharmonic oscillations can be induced in the network by switching operations, temporary earth faults or also by jump resonance. Because of damping in the circuit the amplitude of the oscillation fades away and the natural frequency f_e, depending upon the excitation intensity, can traverse a wide range. Synchronisation of the subharmonic oscillation with the 50 Hz supply frequency oscillation is possible, for example when f_e passes through the value of 25 Hz and both oscillations simultaneously assume a favourable phase relation with respect to each other. The oscillogram of Fig. 3.77 shows subharmonic oscillation of 25 Hz produced in the 2-phase network of Fig. 3.71 after interruption of an earth fault in phase R.

In the same manner as jump resonance, subharmonic oscillations cause overvoltages in the system, which can lead to operational disturbances, e.g., by overloading of the voltage transformers. By connecting resistances to the series connected earth fault indicating windings of single phase voltage

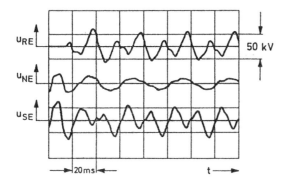

Fig. 3.77 Second harmonic oscillation in a 2-phase network ($f_{1/2} = 25$ Hz)

transformers, the inharmonic oscillations can easily be damped so heavily that they no longer occur.

3.10.2 Experiment

a) Test Setup
The 2-phase network shown in Fig. 3.68 can, for example, be realised with the aid of a test transformer possessing two symmetrical high-voltage windings (Fig.1.1b). The usual adjustable single-phase a.c. supply serves for excitation. As a rule, however, these test transformers are not completely insulated so that their mid-point N must therefore be connected to the low voltage winding at earth potential. But, since just the potential shifts of N are to be investigated in this experiment, these transformers must be energised via a special insulating transformer. The following equipment was used in the course of this experiment :

> T   Test transformer 220 V/ 2x 50 kV, 5 kVA
>   Isolating transformer 220/220 V, 50 kV

Two identical single pole insulated inductive voltage transformers are chosen as high-voltage inductances with non-linear characteristics. The following were used:

$$W_1, W_2 \text{ Voltage transformers } \frac{25000}{\sqrt{3}} \Big/ \frac{100}{\sqrt{3}} \text{ V.}$$

Before the actual experiment, the characteristic $U_{rms} = f(I_{rms})$ of the voltage transformers should be determined, since this will be required for the construction of the diagrams. During these measurements the current must be measured using a true r.m.s. value meter since its shape deviates strongly from the sinusoidal form; an instrument with a moving iron mechanism for example, is suitable.

To reproduce the capacitance $C = C_0 + 2C_{11}$ effective between N and E, the construction elements CB and CM, for example, of the high-voltage construction set can be used. The processes to be investigated develop relatively slowly, so that for the oscillographic measurements a bandwidth of 8 kHz is sufficient. A 4-channel storage type cathode ray oscilloscope was connected to a capacitive voltage divider with CM = 100 pF as the high-voltage capacitor, and to the secondary terminals of the voltage transformer.

b) Earth Fault Overvoltage in a 2-Phase Network

Using the elements described, a circuit of the type shown in Fig.3.68 should be set up. Only CM should be connected between N and E as a concentrated capacitance. For symmetrical operation with a phase voltage $U_{RE} = U_{SE} = 10$ kV, earth faults are simulated by temporarily earthing a phase or by short-circuiting the secondary winding of one of the voltage transformers.

The voltages

$$u_{NE}, u_{RE}, u_{SE} \text{ and } u_{RS}$$

should be oscillographed.

During performance of this experiment, the curves shown in Fig.3.78 resulted for earth fault in phase R. As can be seen, appreciable transient and stationary overvoltages appear at N and S.

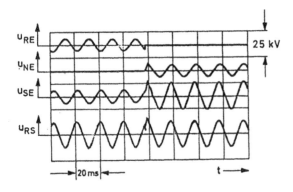

Fig.3.78 Voltage curves for earth fault in phase R in a 2-phase network

c) Jump Resonance in a 2-Phase Network

In the reconstruction of a 2-phase network an additional concentrated capacitance of 1000...3000 pF is connected between N and E. The voltage should be gradually increased under oscilloscopic observation, to the onset of jump resonance.

A voltage lower than that which leads to jump resonance is then applied and jump resonance incited by interruption of the earth fault in a phase.

The oscillograms of Fig.3.79 show the most important curves during the onset of jump resonance. The resonance was caused here by change in the midpoint capacitance.

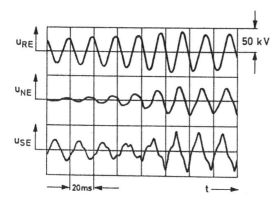

Fig.3.79 Jump resonance in a 2-phase network

d) Jump resonance in a Series Oscillatory Circuit

For these investigations, the 2-phase network is reproduced by a series oscillatory circuit as in Fig.3.72. The circuit used is shown in Fig.3.80. As test transformer T, the instrument denoted by T above in *a)* should be chosen, whereby only one high-voltage winding should be connected. An insulating transformer is no longer necessary. A value of 1000...2000 pF is suitable for the capacitance *C*.

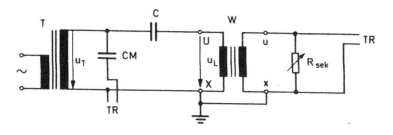

Fig.3.80 Test setup for the measurement of jump resonance and subharmonic oscillations

In this circuit, the transformer voltage and that of the voltage transformer should be determined shortly before and after jump resonance by gradually increasing the excitation. Finally, that value of the loading resistor R_{sec}, connected in the secondary circuit of the voltage transformer with which the jump resonance can be eliminated again, should be found.

e) *Subharmonic Oscillations*

To generate subharmonic oscillations, the earth capacitance C of Fig.3.71 must be substantially increased. With proper choice of C, subharmonic oscillations can be induced by temporarily short-circuiting the secondary winding of the voltage transformer. The transition from the normal operating condition to the resonance stage should be recorded on the transient recorder for both cases. In addition it should be shown that the subharmonic oscillations can be prevented by adequate loading of the secondary winding of the voltage transformer with the resistance R_{sec}.

For this experiment at a transformer voltage of 17 kV and capacitance $C = 13000$ pF, the 25 Hz subharmonic oscillations reproduced in Fig.3.77 were obtained. In the series oscillatory circuit of Fig.3.80, subharmonic oscillations of 16 2/3 Hz could be obtained for $C = 7000$ pF and $U_T = 10$ kV.

3.10.3 Evaluation

The characteristic $U_{rms} = f(I_{rms})$ for the voltage transformer should be drawn on graph paper. The jump resonance voltage U_k should be determined with the aid of the circuit data valid for the experiment 3.10.2d using the graphical construction of Fig. 3.73, and compared with the measured value. In doing so an earth capacitance of 100 ... 300 pF should be taken into account, which is made up of the capacitances of the test transformer, voltage transformer and leads.

The current-voltage ellipse according to Fig. 3.75 should be calculated for the value of the resistance R_{sek} determined under 3.10.2d, drawn into the diagram of a) and briefly discussed.

Example: The determination of the jump resonance voltage U_k is carried out as represented by Fig. 3.73. For this purpose, the single-phase test setup of Fig. 3.80 is converted into the equivalent circuit of Fig. 3.72, applying *Thevenin's* theorem.

From the requirement of equal capacitances, we have:

$$C = C_{11} + C_0$$

and from the requirement of equal open-circuit voltages,

$$U = \frac{C_0}{C_0 + C_{11}} U_\mathrm{T}$$

Using this data the slope of the straight line $(1/\omega C)\, I_\mathrm{rms}$ can be calculated. The point of intersection of the straight line through A_k with the ordinate gives the jump resonance voltage U_k of the equivalent circuit, which, when multiplied by the factor $\dfrac{C_{11} + C_0}{C_0}$ represents the value to be compared to the experimentally determined jump resonance voltage.

The construction of the current-voltage ellipse should also be undertaken along the lines of Fig. 3.75 using the conversion method described above.

With the aid of the potential curves in Fig. 3.77 the stationary potential curves should be constructed for superimposed second subharmonic oscillation at the voltage transformers and at the midpoint capacitances.

Literature: *Rüdenberg* 1953; *Roth* 1959; *Lesch* 1959; *Sirotinski* 1966

3.11 Experiment "Travelling Waves"

A time-dependent variation in the electrical conditions at any point of a spatially extended system is registered by the other parts of the system in the form of electromagnetic travelling waves. If this change occurs in a time of the order of the transit times, the finite propagation velocity must be allowed for. This is always true for networks for energy transmission with long lines when the external or internal overvoltages occur with voltage variations in the range of microseconds to milliseconds. In laboratory practice with extremely large current and voltage variations in the nanosecond range, it is often necessary to consider the spatial setup of the circuit and the equipment, even when spread out over only a few metres, from the point of view of travelling wave theory.

The topics investigated in this experiment by measurements on low-voltage models fall under the following headings:

- Lightning overvoltages,

- Switching overvoltages,

- Surge diverter,

- Protective range,

- Waves in windings,

- Impulse voltage distribution.

It is assumed that the reader has some basic knowledge of the propagation of electromagnetic waves on transmission lines.

3.11.1 Fundamentals

The differential element of a loss-free homogeneous transmission line can be described by its inductance L' and its capacitance C' per unit length. If $u(x, t)$ and $i(x, t)$ represent the voltage and current respectively at point x at time t, the solutions of the differential equations are [e.g. *Unger* 1967]:

$$u(x,t) = u_v (x - vt) + u_r (x + vt)$$
$$Z.i(x,t) = u_v (x - vt) - u_r (x + vt)$$

Here

$$Z = \sqrt{\frac{L'}{C'}} \quad \text{is the surge impedance and}$$

$$v = \sqrt{\frac{1}{L'.C'}} \quad \text{the propagation velocity.}$$

u_v and u_r are travelling waves which travel in the positive or negative x direction with velocity v; their time dependence for loss-free lines is determined by the initial or boundary conditions. The maximum value v can reach is the velocity of light c. As guiding values, we have the following:

for overhead lines	$v \approx 300$ m/μs $\approx c$;	$Z \approx 500\Omega$
for underground cables	$v \approx 150$ m/μs ;	$Z \approx 50\Omega$

Addition of the above solutions gives:

$$u = 2u_v - Z.i.$$

This equation should be used to calculate the voltage u_2 at a point of reflection with input impedance Z_2 at the end of a homogeneous transmission

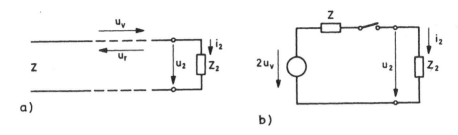

Fig.3.81 Homogenous transmission line terminated by Z_2
a) circuit diagram,   b) travelling wave equivalent circuit

line (Fig. 3.81a). For $u = u_2$ and $i = i_2$, an expression results which is reproduced by the travelling wave equivalent circuit of Fig. 3.81b.

The switch should be closed on arrival of the wave at the point of reflection. If τ_{min} is the lowest transit time occurring in a system for reflections travelling back to the point of reflection, the travelling wave equivalent circuit is valid for times $t \leq 2\ \tau_{min}$.

Introducing the reflection factor $r = u_r / u_v$, we have:

$$u_2 = u_v + u_r = u_v(1+r) = \frac{Z_2}{Z + Z_2} 2u_v$$

$$r = \frac{Z_2 - Z}{Z_2 + Z}$$

The travelling wave equivalent circuit is particularly suitable for the determination of current and voltage at the end of an electrically "long" transmission line, or a line matched at the generator end.

a) Origin of Travelling Waves

Travelling waves as a consequence of lightning discharges. The front-times of the resulting travelling waves lie in the ms range, the tail times are of the order of 100 μs. For a direct stroke to the conductor, the transmission line is suddenly connected to a strong energy source. One may assume that a lightning current i_B is impressed and increases at rates of between 10 and 20 kA/μs. As a result of the lightning currents flowing in, current and voltage waves travel from the point of impact along the conductor.

For the case represented in Fig. 3.82 the voltage appearing at the point of impact is:

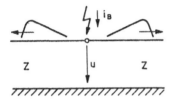

Fig. 3.82 Origin of travelling waves due to a lightning stroke

$$u = \frac{1}{2}Z.i_B$$

For overhead lines, the rate of rise of the resulting overvoltage can be calculated as:

$$S = \frac{du}{dt} = \frac{1}{2}Z\frac{di_B}{dt} = 2.5 \dots 5\ MV / \mu s.$$

50% of all the lightning strokes reach a peak value > 30 kA, and only 10% of all the strokes have a maximum current > 60 kA. The amplitude of these lightning impulse voltages is limited to a value corresponding to the insulation level by flashovers at the insulator chains of the neighbouring masts. This value lies in the region of 2 to 5 times the peak value of the operating voltage.

For a direct stroke to the earth wire or to the mast of an overhead line, as a result of the earth resistance, the mast can temporarily be at such a high potential that a reverse flashover occurs from the mast to one of the conductors. Hence the risk of a reverse flashover occurring is especially large for unfavourable earthing conditions.

Moreover, external overvoltages can also be caused by an indirect stroke. Here the thunder cloud discharges itself in the vicinity of an overhead line in the form of a flash of lightning. The charge, induced on the conductor before the discharge, progresses along the line after the lightning discharge in the form of travelling waves. The amplitude of waves following indirect strokes are comparatively low (up to about 200 kV), but are still dangerous for medium and low-voltage networks and telephone systems.

Travelling waves as a consequence of switching operations. Internal overvoltages which occur as a result of switching operations are of special significance in ultra high-voltage networks. The amplitudes of these switching impulse voltages are only about 2 to 3 times the peak value of the operating voltage. However, since the electrical strength of inhomogeneous electrode configurations for large spacings in air is very low for switching impulse voltages, these largely define the dimensions of air clearances at high nominal voltages (> 400 kV). The front-times are in the region of a few hundred μs and the tail-times in the ms range.

Particularly high voltages can occur during switching operations in connection with short-circuits or earth faults, as well as on switching-off unloaded transformers and capacitances (unloaded cables and overhead lines, capacitor batteries, etc.).

Travelling waves in laboratory work and testing practice. Extremely steep current and voltage variations quite often occur during breakdown mechanisms. Travelling waves are then induced on the conductors and the measuring cables, which can lead to disturbances during measurement and endanger parts of the equipment. Travelling waves also occur on stressing electrical equipment with steep impulse voltages. The potential distribution in spatially extended insulation systems is affected by travelling waves. These phenomena also play a role in high-voltage generators which employ reflection mechanisms to generate high-voltage pulses (see section 1.3.8).

b) Limitation of Overvoltages Using Surge Diverters

The voltage at an installation can be limited with the aid of surge diverters. For high voltages the surge diverters are built up using a series multiple-gap section F and a current-dependent resistance $R(i)$, as shown in Fig. 3.83a. Should the terminal voltage be greater than the breakdown voltage U_A of the gap, it is reduced to the voltage drop across the resistance $U_R = iR(i)$. The terminal voltage u of the surge diverter and the current i

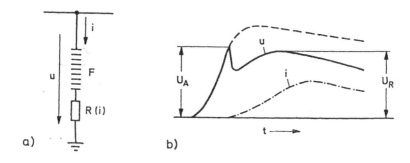

a)   b)

Fig. 3.83 Limitation of overvoltages by valve-type diverters
a) equivalent circuit of a surge diverter
b) time-dependent curves of voltage and current at the surge diverter

for the limitation of an impulse voltage are shown in Fig. 3.83b; for the current curve i the effect of stray inductances has been taken into account. In metal-oxide surge diverters, the sphere-gap is eliminated in view of the large non-linearity of the resistance material.

A surge diverter can guarantee reliable limitation of the voltage U_A in every case at its terminals only. At a certain distance away from the diverter higher voltages may occur. The length of the conductor in front of or behind the surge diverter, within which a definite permissible overvoltage U_{zul} shall not be exceeded for a given waveform, is known as the protective range a. For square waves, the full amplitude is present up to the surge diverter itself; reliable protection against travelling waves from either side is therefore effected only by setting up two surge diverters. The stretch of conductor lying in-between is then fully protected.

In practice, one may assume that travelling waves appear only with a finite voltage steepness S. The protective range for these wedge-shaped waves will be derived on the basis of Fig. 3.84. Let a surge diverter be installed at the point 2 along a homogeneous transmission line; at $t = 0$, the crest of the wave arrives at this position. The distance a between the points 1 and 2 is covered by the wave in time $\tau = a/v$. At $t = U_A / S$ the surge diverter responds to U_A and a backward wave is initiated, the shape of which can, for example, be determined with the aid of the travelling wave equivalent circuit. Assuming a linear increase of the incoming wave and ideal behaviour of the surge diverter, the reflection at point 2 is a wedge-

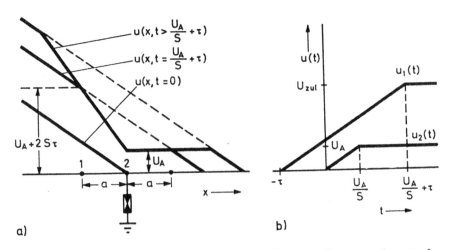

Fig. 3.84 Determination of the protective range of a surge diverter on impact of a wedge-shaped wave
a) $u = f(x)$,   b) $u = f(t)$

shaped wave of slope $- S$. Only after the further interval of the time τ does a voltage limiting effect set in at the point 1, at the time $t = \dfrac{U_A}{S} + \tau$. At this instant the voltage u_1 has a value:

$$u_1 = U_A + 2S\tau$$

Since u_1 is to be $\leq U_{zul}$, for the protective range we have:

$$a = \frac{U_{zul} - U_A}{2S} v.$$

c) Travelling Waves in Transformer Windings

Development of an equivalent circuit. A technically important special case of travelling wave propagation is the impact of an impulse voltage in a transformer winding. Whereas at low frequencies the voltage distribution in the winding is linear because of the magnetic flux linking all the turns, at higher frequencies, as in the spectrum of an impulse voltage, it is also determined by the capacitances. To a first approximation one can perform the calculations using the equivalent circuit of a high-voltage winding as per Fig.3.85, suggested by *K. W. Wagner* in 1915. *L* and *C* denote the inductance and capacitance respectively of an element of the winding and C_e the corresponding earth capacitance.

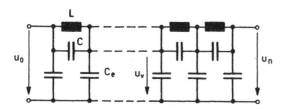

Fig. 3.85 Equivalent circuit of a transformer winding for impulse voltage stress

An impulse voltage $u_o(t)$ applied to the high-voltage terminals will travel with a finite velocity along the individual turns to the end of the winding, be reflected there, and so on. Superimposed on the process however, is a wave which takes a shorter path, namely primarily via the capacitive coupling between individual parts of the winding. The oscillations produced inside the winding in this way result in a time-varying, asymmetrical voltage distribution, which may cause individual parts of the winding to be gravely overstressed .

Calculation and control of the voltage distribution. At the instant of impact of a wave ($t=0$), the capacitances alone determine the voltage distribution in the winding. Hence, for the calculation of the initial distribution for steep waves $L = \infty$ can be assumed. It can be shown that for the simple case of a square wave of amplitude U_0 on the earthed winding ($u_n = 0$) as in Fig. 3.85 the following relationship holds:

$$U_v = U_0 \frac{\sinh(n-v)\alpha}{\sinh n\alpha} \quad \text{with } \alpha = \sqrt{\frac{C_e}{C}}$$

The part of the capacitive equivalent circuit shown in full lines in Fig. 3.86a corresponds to the requirements of the voltage distribution indicated by (C, C_e) in Fig. 3.86b. There is also quite often capacitive coupling to the high-voltage electrode besides that to earth: this is taken into account by the elements C_h in Fig. 3.86a.

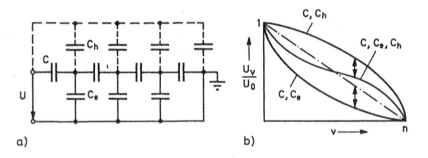

a)   b)

Fig. 3.86 Impulse voltage distribution in a transformer winding
a) equivalent circuit with effective interturn and coupling capacitances
b) voltage distribution on impact of a wave

The capacitive initial voltage distribution, taking into account C_e and C_h at the same time, can no longer be calculated in a straightforward way. For not too great a divergence from the linear distribution, the resultant distribution can be approximately derived from Fig. 3.86b, whereby, for a given v, the amount of deviation from the linear distribution produced by C_e only is deducted from that value effected by C_h alone [*Philippow* 1966].

The capacitances can be varied by appropriate construction of the transformer, and the voltage distribution thereby influenced to advantage. An increase of C relative to C_e effects linearisation. This may be achieved by the chosen type of winding, for example by layer winding or by disc winding [*Kind, Kärner* 1982]. Another possibility is the installation of large-area electrodes (shields) at the high voltage terminals of the winding; these increase the coupling capacitances C_h. The current taken by the earth capacitances then flows partly through the high voltage capacitances C_h.

A linear voltage distribution is obtained in the transformer when for each winding element a current flowing across C_h is equal to that flowing across C_e.

. The smaller the divergence of the initial distribution from the linear final distribution, stressing of the winding in the voltage rise would not only be more uniform, but the amplitudes of the subsequent transient phenomena would also be reduced. Windings with a linear initial distribution are often characterised as non-oscillatory for this reason. Particularly severe stressing occurs during a test with chopped impulse voltages because of the rapid voltage variations which then take place.

3.11.2 Experiment

a) Model investigations at low voltage
Travelling wave investigations of networks, and in particular of transformer windings, are often carried out on models. For this purpose a generator is commonly used which can supply impulse voltages with adjustable waveform and with a peak voltage of a few 100 V at a definite impulse repetition frequency of, say, 50 Hz. The repetitive surge generator[4] RG of this kind is available for this experiment. By synchronising the generator and the oscilloscope, a stationary diagram can be obtained on the screen of the cathode ray oscilloscope (KO).

With these generators, real power transformers can be impulse tested before they are mounted inside the tank in order to determine the impulse voltage distribution; or for design purposes measurements can be made on a model. In a similar manner, it is also possible to simulate transmission lines, for which either cables or four-pole chains may be used.

For convenient realisation of the transmission line model, a coaxial delay cable with $Z = 1500\ \Omega$ and $v = 4.0$ m/μs was chosen for this experiment.

A circuit with a Zener diode is suitable as a model of surge diverter. The Zener voltage of about 30 V corresponds to the residual voltage of the diverter. A double-beam cathode-ray oscilloscope with a bandwidth ≥ 10 MHz should be available for the measurements.

b) Travelling Waves on Transmission Lines
Experimental investigations. An overhead line should be reproduced as in Fig. 3.87 by two cables K1 and K2, each with a transit time of $\tau/2$.

The beginning of the line is terminated by the surge impedance of the cable so that the output terminals 0-0′ of the connected repetitive surge

---

[4] Manufacturer : Haefely-Trench, Basel

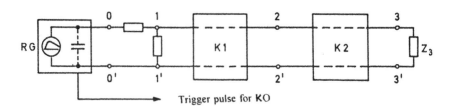

Fig. 3.87 Low voltage model for travelling wave investigations on transmission lines
RG: repetitive surge generator,
K1,K2 : cables with $v = 4m/\mu s$, each 12m long

generator RG can be considered to be short-circuited by a large capacitance.
The generator should be adjusted to deliver a 1.2/50 impulse voltage with
connected load.

The voltage curves at the terminals 1-1´, 2-2´ and 3-3´ should be
oscillographed for various operating conditions. The measurement should
be performed for various terminations of the transmission line at the
terminals 3-3´. As an example, the curves for an open-circuited line
($Z_3 = \infty$) are reproduced in Fig. 3.88 and for a line terminated by a
capacitance in Fig. 3.89.

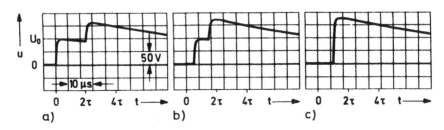

Fig. 3.88 Oscillograms of voltage curves for the transmission line model in Fig. 3.87
for $Z_3 \rightarrow \infty$
a) measurement at the terminal 1-1´ b) measurement at the terminal 2-2´
c) measurement at the terminal 3-3´

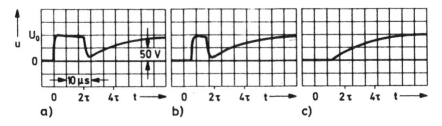

Fig. 3.89 Oscillograms of the voltage curves for capacitive termination with C
a) measurement at the terminal 1-1´ b) measurement at the terminal 2-2´
c) measurement at the terminal 3-3´

Computer Simulation. For calculation of travelling wave phenomena, a large number of suitable commercial software packages is available that can be run on personal computers also, e.g. EMTP(Electromagnetic Transients Programme), SPICE. The investigations to be conducted experimentally as described in the previous section shall thus be simulated on the computer.

The repetitive surge generator is represented by a voltage source u_{RG} with an internal resistance $R_{RG} = Z = 1500\ \Omega$. Let the temporal nature of u_{RG} be rectangular and the amplitude 2. $U_0 = 200$ V (see Fig .3.88a). The matching four-pole 0-0'/1-1' is eliminated. The cables K1 and K2 are represented each time as a loss-less transmission line with a travel time of $\tau/2 = 3\ \mu s$ and a surge impedance of $Z = 1500\ \Omega$. The terminating impedance Z_3 shall be varied. Fig. 3.90 shows as an example, the curve of the voltages for an open end, Fig. 3.91 the curve for an inductive termination ($L = 10$ mH).

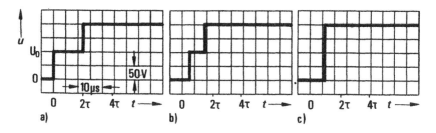

Fig. 3.90 Calculated voltage curves for $Z_3 \to \infty$
a) voltage at the terminals 1-1'   b) voltage at the terminals 2-2'
c) voltage at the terminals 3-3'

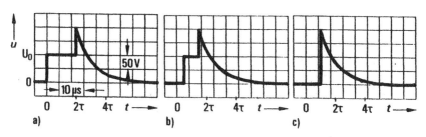

Fig. 3.91 Calculated voltage curves for inductive termination (L = 10mH)
a) voltage at the terminals 1-1'   b) voltage at the terminals 2-2'
c) voltage at the terminals 3-3'

c) Protection by surge diverters

The model of the surge diverter should be connected at position 2-2' in the circuit of Fig. 3.87. The voltage curves should be oscillographed at the terminals 1-1', 2-2' and 3-3', the transmission line being open-circuited ($Z_1 \to \infty$). The curves are reproduced in Fig. 3.92 as an example.

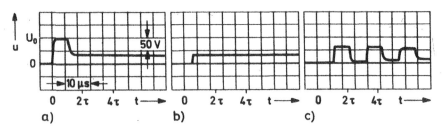

Fig.3.92 Oscillograms of the voltage curves for the installation of surge diverter at position 2-2′ ($Z_3 \rightarrow \infty$)
a) measurement at the terminals 1-1′   b) measurement at the terminals 2-2′
c) measurement at the terminals 3-3′

d) Impulse voltage distribution in transformer windings
For these investigations a transformer model was built with interchangeable high-voltage winding. Thus high-voltage windings with the same number of turns and external dimensions, but with very different properties can be investigated. In the one case the high-voltage winding consists of disc winding (Fig. 3.93a) and in the other case layer winding (Fig. 3.93b).

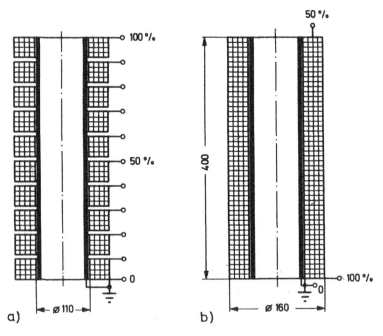

Fig. 3.93 Transformer model for travelling wave investigations
a) high voltage winding in the form of disc winding
b) high voltage winding in the form of layer winding

Both windings are provided with measuring taps at every 10% of the total number of turns (= 1800). The low-voltage winding is connected to earth.

The repetitive surge generator RG should be directly connected to the transformer model. Voltage curves at the various taps should be recorded for different cases:

    I high voltage winding as disc winding (Fig. 3.94)

    II high voltage winding as layer winding (Fig. 3.95)

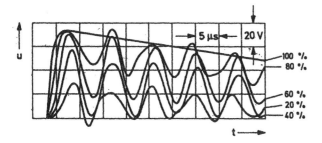

Fig. 3.94 Voltage curves of the transformer model with disc winding during impulse voltage stressing

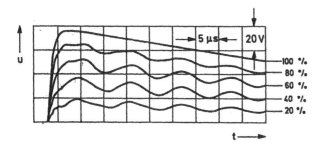

Fig. 3.95 Voltage curves of the transformer model with layer winding during impulse voltage stressing

3.11.3 Evaluation

The lengths of a real transmission line or a real underground table should be determined corresponding to the functional model investigated in 3.11.2b. How large is the ratio of the amplitude of the voltage generated by the repetitive surge generator to that of the voltage entering the line at 1-1'?

For the cases investigated under 3.11.2b the idealized curves for a rectangular shaped voltage and an ideal cable for capacitive termination with C should be drawn and compared with the recorded oscillograms.

For the oscillograms of the measurement according to 3.11.2c with the diverter model idealized curves should be shown and compared with the measured results. The protective range a should be calculated for an assumed permissible voltage $U_{zul} = 40$ V at the line end.

From the oscillograms of 3.11.2d, for the cases I and II, the position, time and amplitude of the highest voltage stress between two measuring points should be determined. For case I the voltage distribution in the transformer winding $u = f(v)$ (with v in %) should be graphically plotted for times $t = 1.5$ μs and $t = 15$ μs (impact of the wave at time $t = 0$) (Fig. 3.96). Where is the maximum stress in the winding for these times?

Literature : *Bewley* 1951; *Strigel* 1955; *Rüdenberg* 1962; *Sirotinski* 1965; *Philippow* 1966; *Heller, Veverka* 1968; *Greenwood* 1971; *Dommel* 1986; *Duyan et al.* 1991

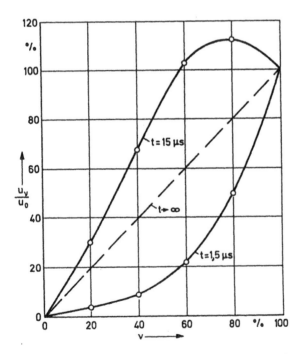

Fig. 3.96 Voltage distribution in a transformer model with disc winding at various times t after impact of a travelling wave.

3.12 Experiment "Impulse Currents and Arcs"

Transient high currents are important to numerous fields of science and technology, because of their magnetic, mechanical and thermal effects. Whenever strong currents coexist with an arc, a heavy concentration of energy results which can also lead to destruction. Lightning currents or the short-circuit currents in high-voltage networks are examples for this. The topics covered by this experiment fall under the following headings:

- Discharge circuit with capacitive energy storage,
- Impulse current measurement,
- Forces in a magnetic field,
- Alternating current arc,
- Arc quenching.

It is assumed that the reader is familiar with section

- 1.4 Generation and measurement of impulse currents.

3.12.1 Fundamentals

a) Force Action in a Magnetic Field
The force action in a magnetic field shall be demonstrated using the transformer arrangement of Fig. 3.97. Position and dimensions of the outer winding 1 carrying a current $i_1(t)$ should be invariable so that only the forces effective on the shorted inner winding 2 are to be observed. Radial forces $\vec{F}_r$ tend to reduce the diameter, and the axial forces $\vec{F}_a$ the length of the inner winding. In a fully symmetrical arrangement ($a = 0$), all the forces acting at the centre of gravity of the inner winding just compensate each other and the system is in a state of unstable equilibrium. On the other hand, if the inner winding is axially displaced ($a \neq 0$), a resultant axial force $\vec{F}$ is produced tending to increase the asymmetry.

The inductance L of the arrangement, measured at the terminals of the outer winding increases with a from the value $a = 0$ to the value for completely removed inner winding. The magnetic energy

$$W = \frac{1}{2} L i_1^2$$

also varies accordingly. For the resultant force acting at the centre of gravity of the inner winding, we have :

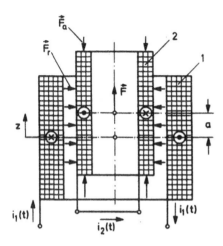

Fig. 3.97 Transformer arrangement
1 primary winding 2 secondary winding

$$F(z) = \frac{dW}{dz} = \frac{1}{2}i_1^2\frac{dL}{dz} + \frac{1}{2}L\frac{d(i_1^2)}{dz}$$

The current pulse is often of such short duration that the reaction of the motion does not affect the magnitude of the force. In this case, since $z = a$, the inductance of the arrangement remains constant during the current interaction period. The initial velocity v_0 of the inner winding with mass m can then be calculated as follows, according to the law of conservation of momentum:

$$v_{0(z=a)} = \frac{1}{m}\int_0^\infty F dt = \frac{1}{2m}\left[\left(\frac{dL}{dz}\right)_{(z=a)}\cdot\int_0^\infty i_1^2\, dt + L.\frac{d}{dz}\left(\int_0^\infty i_1^2\, dt\right)_{(z=a)}\right].$$

The kinetic energy supplied to the inner winding by the current pulse is then:

$$W_{kin} = \frac{1}{2}mv_0^2\ .$$

The reliable control of the mechanical forces in electrical circuits among other things determines the short-circuit strength of transformers. These forces can also be made use of in metal-forming, where the work-piece to be formed takes the place of the shorted inner winding of the transformer model. A further useful application is in electrodynamic drives. Here those arrangements are particularly suitable in which the variation of the inductance due to movement of the object to be accelerated is as large as possible. Even for the case of the transformer arrangement discussed above, a driving action may be observed if the inner winding is replaced by a movable conductor, e.g. a metal tube.

b) Alternating Current Arc

Arcs constitute a type of gas discharge very important to electric power technology (see 3.7.1c). They occur, for example, in switching devices on current switch-off, thereby preventing an abrupt interruption of the current, and with that dangerous overvoltages.

As indicated in the schematic representation of Fig. 3.98, arcs possess two space-charge regions in front of the electrodes, which are described as the anode drop and the cathode drop. In air at normal pressure, their length is of the order of 10^{-6} m and both have an almost constant voltage requirement of about 10 V each. Current transport is essentially realised by the electrons on account of their high mobility. They are created at the cathode by thermo-emission as a consequence of heating of the cathode base by bombarding ions and by field emission.

At large arc lengths s the voltage requirement of the arc column determines the total arc voltage u_h. The arc column consists of plasma in

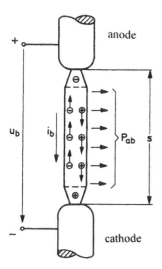

Fig. 3.98 Schematic representation of an arc

which the charge carrier loss, due to recombination and dissipation, is compensated by thermo-ionisation. The column temperature lies in the region from 6000 to 12000 K. Consequently, arcs can only exist when an adequately high energy supply is assured. Stationary arcs require a minimum current of about 1 A and are ignited either by contact breaking or by a breakdown process. If the current supplied by the energy source is insufficient for the existence of an arc, either a glow discharge of appreciably lower current intensity results, or total current interruption.

The power fed into the arc is given by :

$$P_{zu} = i_b \cdot u_b$$

The heat extraction P_{ab} takes place for short arcs via the electrodes, otherwise mainly by thermal conduction, convection and radiation out of the arc column. This can be influenced by the electrode material (carbon, metal) and by the coolant of the arc column (air, oil). If the heat quantity Q stored in the arc is also taken into account, for power equilibrium we have:

$$i_b u_b - P_{ab} = \frac{dQ}{dt} .$$

Neglecting thermal capacitance ($Q = 0$), we obtain the following equation for the arc characteristic:

$$u_b = \frac{P_{ab}}{i_b} .$$

The simple assumption P_{ab} = const is useful in many cases and results in a hyperbolic dependence of the arc voltage upon the current. Cooling of the arc may also result in heat extraction which is approximately proportional to the current; the arc voltage then remains fairly constant. This is the case for example, for very short spacing between metal electrodes since the energy converted in the drop-zones with constant voltage requirement is conducted away through the electrodes.

The arc voltages for both the limiting cases P_{ab} = const and $P_{ab} \sim i_b$ are shown dotted in Fig. 3.99 for the sinusoidal current. A stationary alternating current arc must be re-ignited after each crossover of the current. Moreover, the thermal energy stored in the arc may not be neglected.

The arc voltage requirement therefore, corresponds rather more to the curves drawn in full lines. The voltage value U_Z is known as the ignition peak and that of U_L as the quenching peak.

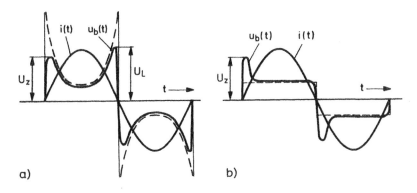

Fig. 3.99 Time variation of the current and voltage of an arc
a) P_{ab} = const., b) $P_{ab} \sim i_b$ (t)

c) Arc Quenching

Whether or not an alternating current arc re-ignites after a current zero depends upon the properties of the arc column and the circuit. The speed of de-ionization of the arc path and the transient recovery voltage which develops along the arc path on interruption of the current, are both critical. The de-ionization process can essentially be influenced by the cooling of the arc. The form of the transient recovery voltage is determined experimentally or calculated by neglecting the effect of the arc [*Slamecka* 1966].

Reliable interruption of a.c. arcs other than in switchgear should also be ensured for the gaps of surge diverters. Here however, in contrast to switchgear, the arc is ignited not by the interruption of a metallic current path but by a breakdown caused by exceeding the breakdown voltage. The a.c. arc so produced is fed from the network mains and must be reliably interrupted by the gap since otherwise the diverter would be overloaded and destroyed.

For effective cooling of the arc, diverter gaps are built up in a series circuit of several gaps with low spacing using electrodes of large area with high cooling efficiency. Particularly good quenching properties are obtained when the arc is additionally cooled by magnetic deflection.

3.12.2 Experiment

a) Measurement of Current in the Discharge Circuit with Capacitive Energy Storage

The experimental circuit is shown in Fig. 3.100. An impulse capacitor C of capacitance 7.5 μF is charged via a charging resistance R_L to a direct voltage U_0, which can be measured across the measuring resistance R_m.

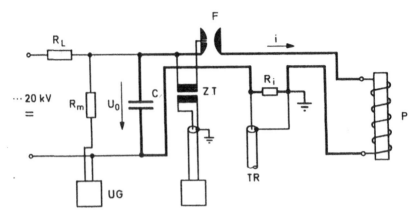

Fig. 3.100   Impulse current circuit with capacitive energy storage

The heavily drawn discharge circuit should be as low-inductive as possible, for example using strip conductors. The trigger gap *F* can be set up as in Fig. 2.17 and is controlled by a trigger generator ZG (see Fig. 2.16) via a transformer ZT, intended for potential separation. The current *i* flowing through the test arrangement P should be measured across the measuring resistance R_i.

Fig. 3.101 shows a coaxially arranged measuring resistor for current measurements in the path of the strip conductor 1; its flat copper conductors are insulated against each other by plastic foil. The current flows through the mounting ring 2 and the screening chamber 3 to the resistance element 4, arranged between the contact discs 5 and 6. It returns to the strip conductor via the threaded connector 7 of the upper contact disc. The two contact discs are separated by the insulating tube 8. The measuring voltage proportional to the current is tapped at the connector terminal 9. To keep the inductance of the measuring resistor at a low value, the spacing between the chamber 3 and the resistance element 4 should be as small as possible.

The testing coil shown in Fig. 3.102 should be investigated as the test setup. The discharge current is fed to the cylindrical primary winding 1, the inductance of which is about 26 μH in a model with the given dimensions and 25 turns. So that the winding is not destroyed by the forces during the experiment, it should be enclosed in an insulating material chamber comprising the parts 2 and 3. The metal tube 4 arranged coaxially with 1 functions as a shorted secondary winding. The tube can be axially displaced on the insulating body 5. Terminals 6 of the primary winding are connected to the strip conductor 8 with the aid of bracing 7.

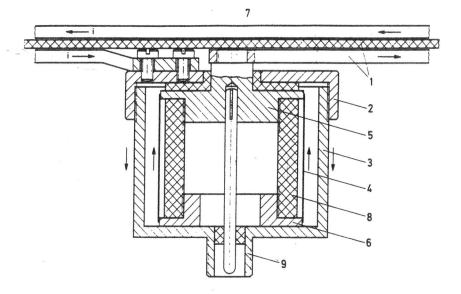

Fig. 3.101 Coaxial measuring resistance in the path of a strip conductor

1 Strip conductor	5 and 6 Contact discs
2 Fixing ring	7 Threaded connector
3 Screening chamber	8 Insulating distance ring
4 Resistance element	9 Connector terminal

The time dependence of the current i should be recorded for the following three cases, at a charging voltage $U_0 = 20$ kV:

- Testing coil P replaced by short-circuiting link,
- Secondary winding 2 of the testing coil in the position shown in Fig. 3.102 ($a = 0$),
- Testing coil without secondary winding

The three oscillograms reproduced in Fig. 3.103 were obtained for this experiment. Since the capacitance C is known, using the periodic time the total inductance of the circuit can be determined and also, by subtraction, that of the individual circuit elements. The following values were obtained:

- Inductance of the setup : 0.238 μH
- Inductance of the experimental arrangement with secondary winding : 11.5 μH.
- Inductance of the experimental arrangement without secondary winding : 25.9 μH.

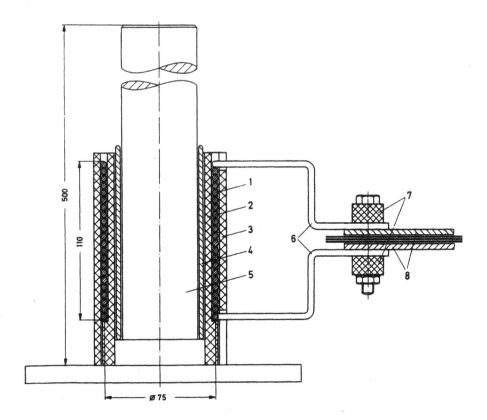

Fig. 3.102 Testing coil for impulse current investigation
1 Primary winding : 25 turns of 6 mm² Cu., 2 and 3 Insulating material chamber
4 Secondary winding, 5 Insulating body 6 Connecting ends 7 Strip conductor
brazing, 8 Strip line conductor

b) Investigation of Electrodynamic Forces
In the same circuit the discharge current is allowed to flow through the
experimental arrangement of Fig. 3.102 with its secondary winding displaced
upwards by $a = 6$ cm using an insulating tube. The secondary winding
then experiences an upward acceleration which results in a certain lift H.
For a suitably chosen charging voltage U_0, the lift obtained may be
determined. At the same time the current oscillogram should be recorded.

For this experiment no frequency variation was obtained over the entire
current curve. From this it follows that the local displacement of the
secondary winding begins practically only after the current has already

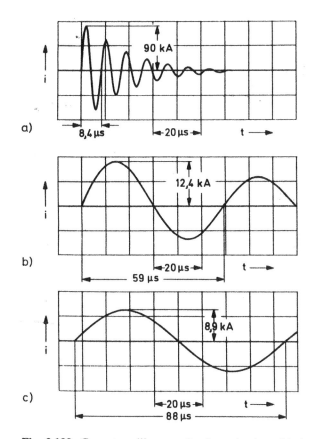

Fig. 3.103 Current oscillograms for determination of inductance
a) Test object short-circuited
b) Secondary winding in central position
c) Without secondary winding

died away ; the conditions for a pulse drive are therefore given. In the investigated arrangement a lift of $H = 20$ cm was obtained for a charging voltage $U_0 = 16.5$ kV. Here the mass of the accelerated tube was 283 g. By equating the potential and kinetic energies an initial velocity $v_0 = 1.98$ m/s can then be calculated.

c) The Stationary A.C. Arc

For this experiment the circuit of Fig. 3.104 should be set up without the parts shown in dotted lines. The a.c. arc should be fed from the low-voltage network. For reasons of safety, an isolating transformer T with ratio 220/220 V should be provided. The experimental setup consists of two

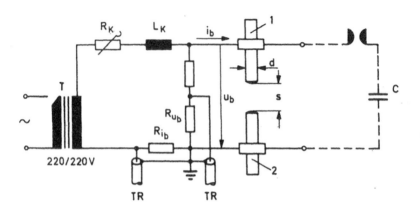

Fig. 3.104 Circuit for investigation of a.c. arcs.
1, 2 Electrodes

cylindrical electrodes 1 and 2 of diameter d. The spacing s can be adjusted using the earthed electrode and an arc discharge struck by contact separation. The current can be adjusted with the inductance $L_K \approx 50$ mH and the resistance $R_K = 0 \ldots 20\ \Omega$, when values up to 10 A should be reached. The arc voltage u_b is measured at the resistance R_{u_b} of a potential divider and the arc current i_b across a measuring resistance R_{i_b}.

Using carbon electrodes with $d = 8$ mm and $s = 5$ mm, u_b and i_b should be recorded as a function of time; the arc is struck by contact separation. By variable fanning of the arc with air and at the same time observing the curve of u_b, the effect of cooling on the arc voltage should be qualitatively observed. Further, the arc characteristic $u_b = f(i_b)$ should be oscillographed.

For this experiment in stationary air the curves shown in Fig. 3.105 were obtained. On fanning the arc, the ignition peak and the quenching peak appeared clearly at first, until finally the arc was extinguished.

d) Operation Test of a Diverter Gap

For these investigations the a.c. arc should be ignited by an impulse current discharge. That part of the circuit shown in dotted lines in Fig. 3.104 should be used for this purpose; it symbolises the circuit for generating impulse currents according to Fig. 3.100. During these experiments the inductance L_K protects the transformer against overvoltages. Carbon electrodes should be used at first. The spacing s should be adjusted so that the breakdown voltage lies below the charging voltage of the capacitor C. The arc initiated

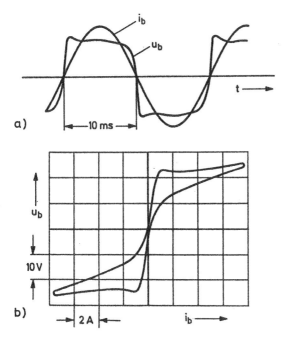

Fig. 3.105 Oscillograms of current and voltage of a stationary a.c. arc, carbon electrodes. s = 5 mm.
a) time dependence   b) dynamic arc characteristic

by the discharge of the capacitor should be observed. The experiment should then be repeated with brass electrodes.

For this experiment reliable striking of the arc by the impulse current circuit was only assured with carbon electrodes. The arc current was about 10 A at a gap spacing of 4 mm. For brass electrodes the electrode spacing had to be reduced to about 0.2 mm in order to obtain a stationary arc at all.

3.12.3 Evaluation

The inductance of the testing coil, with and without the secondary winding, should be determined from the current curves recorded in 3.12. 2a.

The efficiency of the mechanical motion achieved in the investigations under 3.12. 2b should be calculated. To this end, the ratio of the potential energy of the secondary winding to the capacitively stored energy in the impulse capacitor should be derived.

The relationship $P_b = f(t)$ should be constructed from the curves of i_b and u_b recorded under 3.12. 2c.

The different quenching behaviour of carbon and brass electrode arrangements observed in 3.12. 2d should be discussed.

4 Appendix

4.1 Safety Regulations for High Voltage Experiments

Experiments with high-voltages could become particularly hazardous for the participants should the safety precautions be inadequate. To give an idea of the required safety measures, as an example the safety regulations of the High Voltage Institute of the Technical University of Braunschweig shall be described below. These supplement the appropriate safety regulations and as far as possible prevent risks to persons. Strict observance is therefore the duty of every one working in the laboratory. Here any voltage greater than 250 V against earth is understood to be a high voltage (VDE 0100).

Fundamental Rule: Before entering a high-voltage setup every one must convince oneself by personal observation that all the conductors which can assume high potential and lie in the contact zone are earthed, and that all the main leads are interrupted.

Fencing

All high-voltage setups must be protected against unintentional entry to the danger zone. This is appropriately done with the aid of metallic fences. When setting up the fences for voltages up to 1 MV the following minimum clearances to the components at high voltage should not be reduced:

for alternating and direct voltages   50 cm for every 100 kV

for impulse voltages   20 cm for every 100 kV

A minimum clearance of 50 cm shall always be observed, independent of the value and type of voltage. For voltages over 1 MV, in particular for switching impulse voltages, the values quoted could be inadequate; special protective measures must then be introduced.

The fences should be reliably connected with one another conductively, earthed and provided with warning boards inscribed: "High-voltage! Caution! Highly dangerous!" .It is forbidden to introduce conductive objects through the fence while the setup is in use.

Safety-Locking

In high-voltage setups each door must be provided with safety switches; these allow the door to be opened only when all the main leads to the setup are interrupted.

Instead of direct interruption, the safety switches may also operate the no-voltage relay of a power circuit breaker, which, on opening the door, interrupts all the main leads to the setup. These power circuit breakers may only be switched on again when the door is closed. For direct supply from a high-voltage network (e.g. 10 kV city network), the main leads must be interrupted visibly before entry to the setup by an additional open isolating switch.

The switched condition of a setup must be indicated by a red lamp "Setup switched on" and by a green lamp "Setup switched off".

If the fence is interrupted for assembly and dismantling operations on the setup, or during large-scale modifications, all the prescribed precautions for entry to the setup shall be observed. Here particular attention must be paid to the reliable interruption of the main leads. On isolating switches or other disconnecting points, and on the control desk of the setup concerned, warning boards inscribed "Do not switch on! Danger!" must be displayed.

Earthing

A high-voltage setup may be entered only when all the parts which can assume high-voltage in the contact zone are earthed. Earthing may only be effected by a conductor earthed inside the fence. Fixing the earthing leads onto the parts to be earthed should be done with the aid of insulating rods. Earthing switches with a clearly visible operating position, are also permissible. In high-power setups with direct supply from the high-voltage network, earthing is achieved by earthing isolators. Earthing may only follow after switching the current source off, and may be removed only when there is no longer anyone present within the fence or if the setup is vacated after removal of the earth. All metallic parts of the setup which do not carry potential during normal service must be earthed reliably and with adequate cross-section of at least 1.5 mm² Cu. In test setups with direct supply from the high-voltage network, the earth connections must be made with particular considerations of the dynamic forces which can arise.

Circuit and Test Setup

Inasmuch as the setup is not supplied from ready wired desks, clearly marked isolating switches must be provided in all leads to the low-voltage circuits of high-voltage transformers and arranged at an easily identifiable position outside the fence. These must be opened before earthing and before entering the setup.

All leads must be laid so that there are no loosely hanging ends. Low-voltage leads which can assume high potentials during breakdown or flashovers and lead out of the fenced area, e.g. measuring cable, control cable, supply cable, must be laid inside the setup in earthed sleeves. All components of the setup must be either rigidly fixed or suspended so that they cannot topple during operation or be pulled down by the leads. For all setups intended for research purposes, a circuit diagram shall be fixed outside the fence in a clearly visible position.

A test setup may be put into operation only after the circuit has been checked and permission to begin work given by an authorised person.

Conducting the Experiments

Everyone carrying out experiments in the laboratory is personally responsible for the setup placed at his disposal and for the experiments performed with it. For experiments during working hours one should try, in the interest of personal safety, to make sure that a second person is present in the testing room. If this is not possible, then at least the times of the beginning and ending of an experiment should be communicated to a second person.

When working with high-voltages beyond working hours, a second person familiar with the experimental setups must be present in the same room.

If several persons are working with the same setup, they must all know who is to perform the switching operations for a particular experiment. Before switching on high-voltage setups, warning should be given either by short horn signals or by the call "Attention! Switching-on!". This is especially important during loud experiments, so that people standing-by may cover their ears. If necessary, switching off can be announced after completion either by a single long hooting tone or by the call: "Switched off".

Explosion and Fire Risk, Radiation Protection

In experiments with oil and other easily inflammable materials, special care is necessary owing to the danger of explosion and fire. In each room where work is carried out with these materials, suitable fire extinguishers must be to hand, ready for use. Easily inflammable waste products, e.g.

paper or used cotton waste, should always be disposed off immediately in metal bins. Special regulations must be observed when radioactive sources are used.

Accident Insurance
Everyone working in the Institute must be insured against accidents.

Conduct During Accidents
Mode of action in case of an electrical accident:

1. Switch off the setup on all poles. So long as this has not been done, the victim of the accident should not be touched under any circumstances.
2. If the victim is unconscious, notify the life-saving service at once. Telephone .... Immediate attempts to restore respiration by artificial respiration or chest massage! These measures must be continued, if necessary, up to the beginning of an operation.(Only 6 to 8 minutes′ time before direct heart massage!).
3. Even during accidents with no unconsciousness, it is recommended that the victim lies quietly and a doctor's advice be sought.

4.2 Calculation of the Short-Circuit Impedance of Transformers in Cascade Connection

In the general case, the stages of a cascade according to Fig. 1.2 consist of three windings, the potentials of which are independent of one another. This condition is fulfilled by the equivalent circuit in Fig. 4.1, where an impedance $\underline{Z}_E$, $\underline{Z}_H$ or $\underline{Z}_K$ is attributed to each winding; each impedance is connected in series with an ideal 3-winding transformer with the corresponding number of turns N_E, N_H or N_K. The impedances are determined either from calculated or experimentally derived results of three short-circuit tests between any two windings taken at a time [e.g., *Siemens* 1960].

For each stage it follows that when the magnetizing current is neglected, the sum of ampere-turns of all the windings must be equal to zero:

$$N_E \cdot I_E - N_H \cdot I_H - N_K \cdot I_K = 0.$$

The method of calculation shall be illustrated on the example of a 3-stage cascade, where the losses shall be neglected for the sake of clarity:

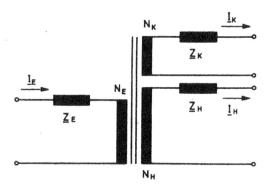

Fig. 4.1 Equivalent circuit of one stage of a cascade

$$\underline{Z}_E = j \cdot X_E \; ; \; \underline{Z}_H = j \cdot X_H \; ; \; \underline{Z}_K = j \cdot X_K$$

Further, it will be assumed that the ratio of the number of turns is the same for all stages, viz:

$$\frac{N_E}{N_H} = \frac{N_K}{N_H}$$

The result of the assumptions made above is the equivalent circuit shown in Fig 4.2a . The indicated currents and the dashed reactances refer to the number of turns N_H of the respective high-voltage winding.

An equivalent circuit as in Fig. 4.2b shall now be derived for the entire cascade. The resulting short-circuit reactance X_{res} is obtained from the condition that the power rating be the same :

$$I_H^2 \, X_{res} = \sum_{v=1}^{3} \left(I_{Ev}'^2 \, X_{Ev}' + I_{Kv}'^2 \, X_{Kv}' + I_{Hv}^2 \, X_{Hv} \right).$$

From this it follows at once that:

$$X_{res} = X_{H1} + X_{H2} + X_{H3} + X_{K2}' + X_{E3}' + 4(X_{K1}' + X_{E2}') + 9X_{E1}'.$$

With the simplifications as before, one obtains for the short-circuit reactance of an n-stage cascade, with $X_{Ko}' = 0$:

$$X_{res} = \sum_{v=1}^{3} \left[X_{Hv} + v^2 \left(X_{E(n+1-v)}' + X_{K(n-v)}' \right) \right].$$

This method is not bound to the simplifications made here for the sake of clarity; it can easily be extended to different transformer ratios and can

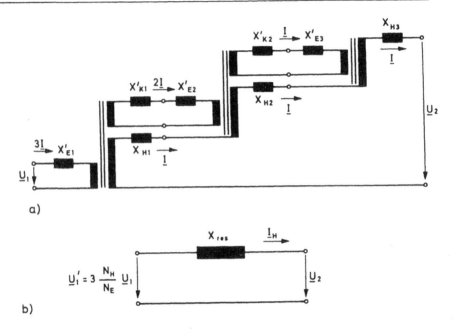

a)

b)

Fig. 4.2 Equivalent circuits of a 3-stage cascade
a) complete equivalent circuit   b) simplified equivalent circuit

also take the effective resistances into account. It may also be used for the calculation of the short-circuit impedance of potential transformers in cascade connection.

4.3 Calculation of Single-Stage Impulse Voltage Circuits

For the circuit b as in Fig. 1.36 the following equations are valid, using the same notations:

$$U_0 - \frac{1}{C}\int_0^t (i_e + i_d)\, dt = i_e R_e = i_d R_d + u(t)$$

$$i_e = C_b \frac{du(t)}{dt} \qquad \text{with } u(t = 0) = 0$$

This differential equation will be solved by applying the Laplace transformation. For the functions in the p-plane, the corresponding capital letters will be used as symbols:

$$\frac{U_0}{p} - \frac{1}{pC_s}\left[I_e + pC_bU\right] = I_eR_e = U\left(pR_dC_b + 1\right)$$

Solving for $U = U(p)$, we have:

$$U(p) = \frac{U_0}{R_dC_b} \cdot \frac{1}{p^2 + bp + c}$$

wherein

$$b = \frac{1}{R_eC_s} + \frac{1}{R_dC_s} + \frac{1}{R_dC_b}$$

$$c = \frac{1}{R_dC_bR_eC_s}$$

The two roots of the quadratic equation in the denominator polynomial are:

$$p_{1,2} = \frac{b}{2}\left(-1 \pm \sqrt{1 - \frac{4c}{b^2}}\right).$$

These are always < 0 and real. Reverse transformation into the t-plane gives:

$$u(t) = \frac{U_0}{R_dC_b} \cdot \frac{1}{p_1 - p_2}\left(e^{p_1t} - e^{p_2t}\right)$$

$$u(t) = \frac{U_0}{R_dC_b} \cdot \frac{\tau_1\tau_2}{\tau_1 - \tau_2}\left(e^{-t/\tau_1} - e^{-t/\tau_2}\right).$$

Here the time constants $\tau_1 = -1/p_1$ and $\tau_2 = -1/p_2$ have been introduced. The general solution can be appreciably simplified if the usually valid approximation

$$R_e \cdot C_s \gg R_d \cdot C_b$$

is considered. Then the relationships

$$b \approx \frac{1}{R_d}\left(\frac{C_s + C_b}{C_sC_b}\right) \quad \text{and} \quad \frac{4c}{b^2} \ll 1$$

follow. With that the square root expression in $p_{1,2}$ approaches the value

$\left(1 - \dfrac{2c}{b^2}\right)$ and it follows that

$$p_1 \approx -\frac{c}{b} = -\frac{1}{R_e(C_s + C_b)}; \qquad \tau_1 = R_e(C_s + C_b)$$

and

$$p_2 \approx \frac{c}{b} - b \approx -b = -\frac{1}{R_d}\left(\frac{C_s + C_b}{C_s C_b}\right); \quad \tau_2 = R_d\left(\frac{C_s C_b}{C_s + C_b}\right).$$

4.4 Calculation of Impulse Current Circuits

For the circuit of Fig. 1.59, the following equation is valid for $t > 0$, with the notations there and with $i = i(t)$,

$$U_0 - \frac{1}{C}\int_0^t i\, dt = R.i + L.\frac{di}{dt}.$$

From this we have the differential equation

$$\frac{d^2 i}{di^2} + \frac{R}{L} \cdot \frac{di}{dt} + \frac{1}{L \cdot C} = 0.$$

Depending on the magnitude of R, the following three solutions are obtained:

the periodic solution for $\qquad\qquad 0 < R < 2.\sqrt{\dfrac{L}{C}}$

the aperiodic border case for $\qquad R = 2.\sqrt{\dfrac{L}{C}}$

the aperiodic solution for $\qquad\qquad R > 2.\sqrt{\dfrac{L}{C}}$

In Table 4.1, the solutions of the differential equation and the characteristic parameters are summarised [*Zischank 1983*]:

Table 4.1

	periodic	border case	aperiodic
$i(t)$	$i = I_0 \cdot \sin\omega t \cdot e^{-t/\tau}$	$i = \frac{U_0}{L} \cdot t \cdot e^{-t/\tau}$	$i = I_0 \left(e^{-t/\tau_1} - e^{-t/\tau_2}\right)$
	with $I_0 = \dfrac{U_0}{\sqrt{\frac{L}{C}-\left(\frac{R}{2}\right)^2}}$ $\tau = \frac{2L}{R}$ $\omega = \sqrt{\frac{1}{LC}-\left(\frac{R}{2L}\right)^2}$	with $\tau = \frac{2L}{R}$	with $I_0 = \dfrac{U_0}{\sqrt{R^2-\frac{4L}{C}}}$ $\tau_1 = \frac{1}{\frac{R}{2L}-K}$, $\tau_2 = \frac{1}{\frac{R}{2L}+K}$ $K = \sqrt{\left(\frac{R}{2L}\right)^2 - \frac{1}{LC}}$
$\left(\frac{di}{dt}\right)_{max}$	$\frac{U_0}{L}$	$\frac{U_0}{L}$	$\frac{U_0}{L}$
$\hat{I}$	$\hat{I} = I_0 \cdot \sin(\omega T_p) \cdot e^{-T_p/\tau}$	$\hat{I} = \frac{2}{e} \cdot \frac{U_0}{L}$	$\hat{I} = I_0 \cdot \left(e^{-T_p/\tau_1} - e^{-T_p/\tau_2}\right)$
T_p	$T_p = \frac{1}{\omega} \cdot \arctan(\omega\tau)$	$T_p = \tau$	$T_p = \frac{\tau_1 \cdot \tau_2}{\tau_1-\tau_2} \cdot \ln\frac{\tau_1}{\tau_2}$
$\int i\,dt$	$\frac{I_0\cdot\omega}{\omega^2+\frac{1}{\tau^2}}\left[\frac{2}{1-e^{-\pi/(\omega\tau)}}-1\right]$	$U_0 \cdot C$	$U_0 \cdot C$
$\int\limits_0^{\infty} i^2\,dt$	$\frac{I_0^2}{4}\cdot\frac{\tau}{1+\left(\frac{1}{\omega\tau}\right)^2}$	$\frac{U_0^2\cdot C}{4}\sqrt{\frac{C}{L}}$	$\frac{I_0^2}{2}\cdot\frac{(\tau_1-\tau_2)^2}{\tau_1+\tau_2}$

4.5   Calculation of the Impedance of Plane Conductors

In high-voltage setups rapidly varying high currents are often passed through extended plane conductors. In choosing the dimensions of these conductor systems the problem arises of determining the voltage drops which occur or the impedances responsible for them. To do this however, it is first necessary to reach an agreement about what shall be considered an impedance. This shall be done on the basis of Fig. 4.3.

As an example of an open system as in a) consider a conductor band formed into a circular cylinder carrying an alternating current $\underline{I}$, as shown by the indicated arrows. On account of the enclosed magnetic flux a voltage $\underline{U}_1$, would be measurable across the shortest distance between points 1 and 2 even if the band were an ideal conductor. Decisive for the impedance of the conductor band can therefore sensibly be only that voltage $\underline{U}$ which would be measured by a circumferential arrangement of the voltage measuring leads:

$$Z = \frac{U}{I}$$

Moreover, for practical cases, U can also be considered as that voltage by which the voltage U_1 is greater than in the case of an ideal conductor.

In closed systems, such as the cylindrical chamber in Fig. 4.3b, no magnetic field occurs on the outside, which is why the arrangement of the voltage measuring leads to determine the voltage U between points 1 and 2 can be arbitrary. Nevertheless, it should be observed here that for high angular frequencies $\omega = 2\pi f$, the current density on the external surface of plate-type plane conductors, consisting of metallic foils for example, will be very low, and then no potential difference would be measurable from the outside. This situation occurs when the depth of penetration

$$\delta = \frac{1}{\sqrt{\pi\mu\kappa f}}$$

is small compared with the thickness w of the plate-type plane conductor. Here $\mu = \mu_0 \cdot \mu_r$ denotes the permeability and κ the specific conductivity. The following are some guiding values for copper and iron ($\mu_r = 200$):

$f = 1$ MHz :	$\delta_{Cu} = 0.07$ mm,	$\delta_{Fe} = 0.021$ mm
$f = 100$ kHz :	$\delta_{Cu} = 0.21$ mm,	$\delta_{Fe} = 0.04$ mm

An application of the system as shown in Fig. 4.3b is the measurement of the voltage drop in the walls of a completely screened laboratory. Fig. 4.3c shows a system which also has a hollow cylindrical conductor, as used in foil measuring resistors.

From a current-carrying plane conductor a small square can always be cut such that the current paths on two opposite sides are parallel. Its impedance is defined as the specific plane impedance :

$$\underline{Z}' = R' + j\omega L'$$

It is identical to the impedance of a square plane conductor of arbitrary length carrying parallel component currents, as would be the case for $l = b$ in Fig. 4.4. This is true for the plate-type plane conductor shown under a) as well as for the lattice-type shown under b). In the latter case, the measured

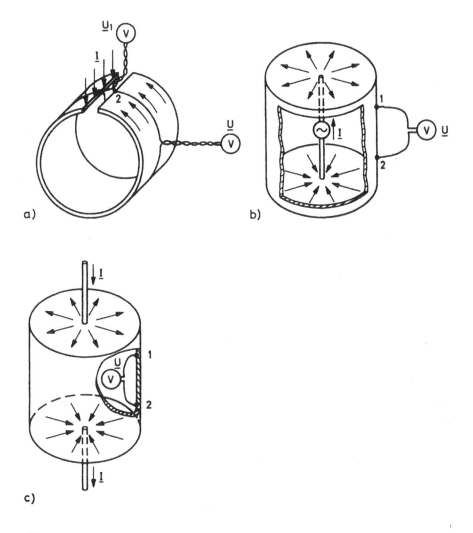

Fig. 4.3 Models for defining the impedance of plane conductors
a) open system
b) system with coaxial internal return
c) system with external return

result is independent of the angle β between the lattice bars and the sides of the square, so long as the spacing between the component conductors is $a \ll b$.

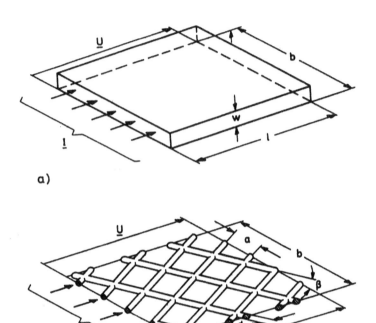

a)

b)

Fig. 4.4 Plane conductor with parallel component current direction
a) plate-type plane conductor
b) lattice-type plane conductor

For the specific plane impedances of the most important types of conductors, the relationships summarized below are valid for plate-type plane conductors of thickness w [*Lautz* 1969]:

for $w \ll \delta$ it follows that $R' = \dfrac{1}{\kappa w}$ $;L' = \dfrac{\mu w}{8}$.

for $w \gg \delta$, assuming double-sided current flow, it follows that

$$R' = \frac{1}{\kappa 2\delta}; \quad L' = \frac{\mu 2\delta}{8} = \frac{R'}{\omega}$$

For single-sided current flow the effective thickness of the conducting layer is halved and therefore

$$R' = \frac{1}{\kappa\delta}; \quad L' = \frac{\mu\delta}{2} = \frac{R'}{\omega}$$

Lattice-type plane conductors with wire diameter d and mean spacing of wires $a \gg d$ [*Sirait* 1967 ; *Hylten-Cavallius, Giao* 1969]:

for $d \ll \delta$, we have

$$R' = \frac{4a}{\kappa\pi d^2}; \quad L' = L'_i + L'_a$$

$$L'_i = \frac{\mu a}{8\pi}$$

$$L'_a = \frac{\mu_o a}{2\pi} \ln\left(\sin\frac{\pi d}{2a} \right),$$

for $d \gg \delta$ it follows

$$R' = \frac{a}{\kappa\pi d\delta}; \quad L' = L'_i + L'_a$$

$$L'_i = \frac{1}{\omega}\cdot\frac{a}{\kappa\pi d\delta} = \frac{R'}{\omega}$$

L'_a as for $d \ll \delta$.

Under the condition that $a \gg d$, these relationships are also valid when instead of the lattice-type plane conductor with square lattice shown in Fig. 4.4b, other forms are considered, such as hexagonal lattices.

For plane conductors with parallel component currents, the impedance of a suitably chosen rectangle of breadth b and length l as in Fig. 4.4 is given by:

$$\underline{Z} = \underline{Z}'\frac{l}{b} \cdot$$

For plane conductors with radial component currents the impedance of a suitably chosen circular ring with radii r and R as in Fig. 4.5 can be calculated as:

$$\underline{Z} = \underline{Z}'\frac{1}{2\pi} \ln\frac{R}{r} \cdot$$

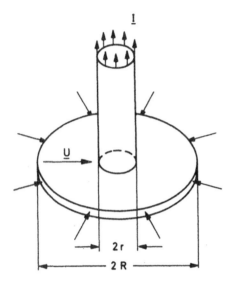

Fig. 4.5 Plane conductor with a single current junction

It is often required to know the potential difference between two current junctions in an extended plane conductor. Such an arrangement is shown in Fig. 4.6, where it is assumed that the currents are led into the plane conductor by cylindrical conductors of radius r at a mean spacing c ; the

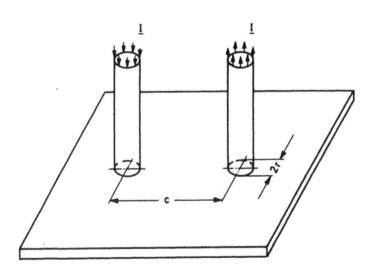

Fig. 4.6 Plane conductor with double current junction

dimensions of the plane conductor are very large compared with c. Analogous to the capacitance of a double line [e.g. Lautz 1969], we have:

$$Z = Z'_- \frac{1}{\pi} \ln \left(x + \sqrt{x^2 - 1} \right) \quad \text{with} \quad x = c/2r.$$

4.6 Statistical Evaluation of Measured Results

For the experimental determination of the electric strength of insulation systems, one obtains measured values which can show appreciable dispersion, depending upon the insulating material and the electrode configuration. If the fluctuations of the measured values are of a random nature, then it is appropriate to apply the methods of mathematical statistics to their evaluation[1]. In this way it is possible to make statements of reliable certainty about the performance of a large assembly with the aid of only a few measurements. Moreover, the results can be represented in a simple and clear manner and easily compared with one another. In certain cases, using statistical methods, it is possible to show that different mechanisms are at play, for instance when a series of measurements may be divided into various subgroups.

The application of statistics shall be discussed here on the example of the breakdown discharge voltage U_d, which is especially important in high-voltage technology; the same principles apply to other measured quantities. It is useful to distinguish between two groups of results; these shall be treated below under 4.6.1 and 4.6.2.

4.6.1 Direct Determination of Probability Values (Series Stressing)

In a first group of investigations a voltage of given time dependence is repeatedly applied to the same sample (or to several identical samples for destructive breakdown discharges); the number of breakdown discharges n_d out of a total number of applications n is determined each time for a specified value of the voltage U. The breakdown discharge probability $P(U)$ = n_d/n is thus directly obtained. For example, when testing insulators with full impulse voltages, the distribution function of the breakdown discharge

---

[1] This appendix has been compiled based on IEC-Publication 60-1 (1989): High-Voltage Test Techniques, Part 1: General definitions and test requirements or VDE 0432-1; Further references, among others, in *Kreyszig* 1967; DIN 1319; *Rasquin* 1972; *Hauschild, Mosch* 1984.

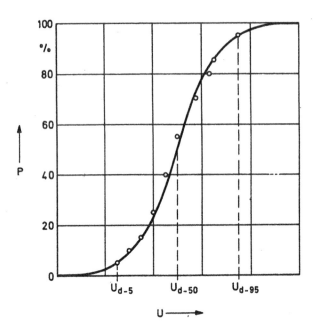

Fig. 4.7 Experimental distribution function for breakdown discharge voltages, plotted in a linear coordinate system.

voltage shown in Fig. 4.7 is directly obtained. Some important characteristic values are the voltages U_{d-50}, U_{d-5} and U_{d-95}. For the evaluation, the measured values of breakdown discharge probabilities for different voltages are suitably plotted in a probability net and one obtains the result shown in Fig. 4.8. If the measured points lie approximately on a straight line as indicated in the figure, it may be assumed that the breakdown discharge voltage of the investigated sample obeys a Gaussian normal distribution law. That is to say the ordinate of this probability net is so divided that the cumulative frequency curve of a normal distribution becomes a straight line. The assumption of a normal distribution for the breakdown discharge voltage of electrode arrangements with gaseous, liquid or solid insulation is permissible in most cases, provided one restricts oneself to the range of about 5...95% breakdown discharge probabilities; outside this range special methods must be adopted [see, for example, IEC Publication 60-1(1989); High Voltage Test Techniques, General Definitions and Test Requirements].

Once the straight line approximating the measured points has been drawn in the probability net, the value $U_d \approx U_{d-50}$ is read off at the breakdown discharge probability $P(U) = 50\%$. Further, the standard deviation s of the measured series is obtained as the difference of the voltages at $P(U) =$ 50% and 16%, or also 50% and 84%, since the Gaussian distribution is symmetrical.

4.6.2 Determination of the Distribution Function of a Measured Quantity

In a second group of investigations, a certain voltage is applied to a sample and increased until a breakdown or flashover occurs. In a subsequent experiment on the same sample (or a new identical sample in case of destructive breakdown discharges) a slightly different value of the breakdown discharge voltage results. Thus one obtains a series of measured U_d-values which show some dispersion. Recording the impulse voltage-

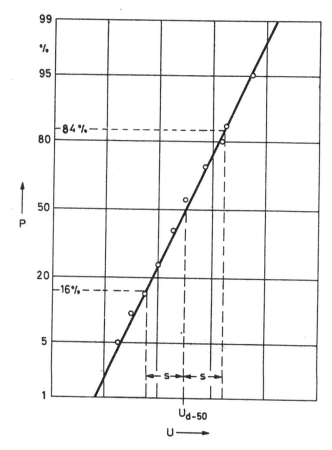

Fig. 4.8 The distribution function of Fig. 4.7 in the probability net

time characteristics of gaps or surge diverters belongs to this type of test for example.

For a series of n breakdown discharge values U_d, the mean value $\overline{U}_d$ and the standard deviation s are calculated using the following equations:

$$\overline{U}_d = \frac{1}{n}\sum_{i=1}^{n} U_{d_i}$$

$$s = \sqrt{\frac{1}{n-1}\sum_{i=1}^{n}\left(U_{d_i} - \overline{U}_d\right)^2}.$$

The standard deviation can also be referred to $\overline{U}_d$ and is then termed the coefficient of variation v:

$$v = \frac{s}{\overline{U}_d}.$$

This calculation can be performed quite generally for any arbitrary distribution. When a normal distribution is assumed, 84% - 16% = 68% of all the U_d values must lie within the range $\overline{U}_d - s$ to $\overline{U}_d + s$.

Alternatively, graphical evaluation of the series on probability paper is also possible, analogous to Fig. 4.8. Here $P(U_d) = n_i / (n+1)$ is plotted in the probability net as a function of U_{d_i} where the breakdown discharge values are arranged according to their magnitude; n_i is the number of disruptive discharges up to and including the voltage U_d, out of a total of n voltage applications. The distribution function is again approximated in the probability net by a straight line, from which $\overline{U}_d$ and s can be determined as described above. Exactly the same values as calculated from the equations are obtained only in exceptional cases; .nevertheless, the graphical method provides a picture of the distribution function which can be extremely informative in many cases.

4.6.3   Determination of the Confidence Limits of the Mean Value of the Breakdown Discharge Voltage

The values of U_d and s, obtained as in 4.6.2 from the limited number of n measurements of a series, in the mathematical sense represent more or less accurate estimates of the corresponding values of the very large total population of samples. Once again, under the assumption of a normal distribution of U_d values, one can specify the limits for a measurement within which the mean value of a series with $n \to \infty$ can be expected to lie for a given statistical certainty P. The calculation of these " mean value confidence limits" is very useful, particularly for the comparison of different series of measurements.

Consider a random sample of n measured values, the mean value and standard deviation of which were calculated as U_d and s respectively. The mean value of the breakdown discharge voltage, determined from an infinitely large number of individual measurements, would then lie for a given certainty P, within the confidence limits of the mean value of the random sample comprising n test samples, viz:

$$\overline{U}_d \pm \frac{t}{\sqrt{n}} \cdot s$$

The factor t depends upon the value chosen for P as well as on the random sample number n and is tabulated in statistical handbooks [Owen 1962; Kreyszig 1967]. For a statistical certainty of $P = 95\%$, the following values may be quoted:

n	5	10	20	50	100	200	∞
t	2.8	2.3	2.1	2.0	2.0	1.97	1.96
$t/\sqrt{n}$	1.24	0.72	0.47	0.28	0.2	0.14	0

Should one determine, on the basis of random sampling, for example, which of the two slightly different types of test samples has the higher electrical strength, then the mean value U_d and its confidence limits are calculated for each random sample. If the confidence intervals of both samples do not overlap, one may then assert that, for the chosen statistical certainty of, for instance, $P = 95\%$, the one model has a higher breakdown discharge voltage. This is represented graphically in Fig. 4.9. If the confidence intervals overlap by more than a quarter of the smaller interval, then the measured difference could be incidental [Sachs 1970]. An example for this is shown in Fig. 4.10.

4.6.4   Details for the Determination of Breakdown Discharge Voltages with a Given Probability

If the mean value and standard deviation of breakdown discharge voltages of an electrode configuration are known, statements about the probable distribution of the measured values can be made, once again assuming a normal distribution. Out of $n = 1000$ independent individual measurements,

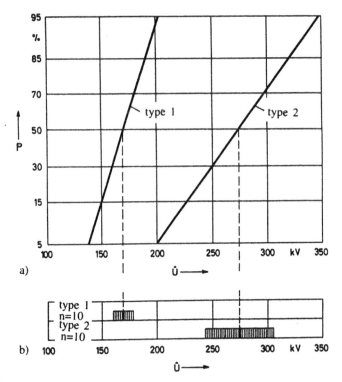

Fig. 4.9 Impulse breakdown discharge voltages of two different types of a test object
a) distribution functions
b) confidence limits of mean values $U_{d\text{-}50}$ ($P=95\%$); an effect of the design on $U_{d\text{-}50}$ is
statistically ensured

317 fall outside the range	$\overline{U}_d \pm s$
46 fall outside the range	$\overline{U}_d \pm 2s$
3 fall outside the range	$\overline{U}_d \pm 3s$

where it is assumed that for $n = 1000$, the measured values of U_d and s differ only minimally from the corresponding values for the total population. To be precise, and particularly for a small random sample number, the confidence limits of U_d and s must be taken into consideration.

In practical measurements, as in 4.6.1, the value $U_d - 3s$ is often used as the estimated value for the impulse withstand voltage $U_{d\text{-}0}$ of an electrode configuration; $U_d + 3s$ is then the assured breakdown discharge voltage $U_{d\text{-}100}$. In the case of a normal distribution of the breakdown discharge

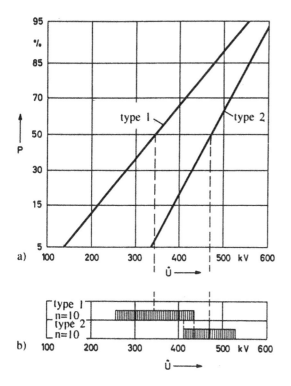

Fig. 4.10 Impulse breakdown discharge voltages of two different types of a test object
a) distribution functions
b) confidence limits of mean values $U_{d\text{-}50}$ ($P=95\%$); no statistically ensured effect of the design on $U_{d\text{-}50}$

voltages, and with sufficiently accurate values of U_d and s, these limiting values correspond to a breakdown discharge probability of 0.14 or 99.86 % respectively.

4.6.5 "Up and Down" Method for Determining the 50% Breakdown Discharge Voltage

When only the breakdown discharge voltage of an electrode configuration, as discussed under 4.6.1, is to be determined with a minimum sacrifice of time and yet to good accuracy, the "up and down" method is especially well-suited for this purpose. With only a small number of measurements this method supplies a very good estimate of $U_{d\text{-}50}$.

Initially one chooses a voltage U_k (an estimated value of the required breakdown discharge voltage), and a voltage interval ΔU_k which should be about 3% of U_k. An impulse voltage with a peak value U_k is then applied to the sample. If no breakdown or flashover occurs, the next impulse is given the peak value $U_k + \Delta U_K$. If a breakdown discharge occurs, the next peak value is $U_k - \Delta U_k$. This process is continued, whereby the peak value of each impulse voltage is determined by the result of the preceding experiment. The number of impulse voltages n_i for every peak value U_i obtained as in the above procedure is recorded; the 50% breakdown discharge voltage can then be determined using the following equation:

$$U_{d-50} = \frac{\sum n_i U_i}{\sum n_i} .$$

Indeed, even for $\sum n_i = 20$, the value so determined lies, to a high degree of certainty, within the range of breakdown discharge probability between $P(U) = 30\%$ and $P(U) = 70\%$.

The standard deviation can also be obtained from this kind of series for determining U_{d-50} [Dixon, Massey 1969]. However, a large number of measurements $\sum n_i$ would then be necessary.

4.7 Specifications for High Voltage Test Techniques

A specially responsible job of the engineer of high voltage technology is conducting acceptance tests. Thereby, it is important for the manufacturer and the user of high voltage equipment to have at their disposal an up-to-date from the point of Science and Technology and binding guidelines to the conduct of the tests. Specifications have therefore a great practical significance in this area.

Since high-voltage insulating systems and equipment are, as a rule, offered in a world wide market, international specifications are specially required for this purpose. This task is undertaken by the Technical committee TC 42 "High-Voltage Testing Techniques" of the International Electrotechnical Commission (IEC); the appropriate national standards agree, in general, contentwise fully with the international specifications.

At the moment, the following IEC specifications are available in the area of high voltage testing techniques(the corresponding DIN VDE standards are given within brackets):

IEC Publ. 60-1   High-voltage test techniques
                  Part 1 : General definitions and test requirements, 1989
                  (DIN VDE 0432-1)
IEC Publ. 60-2   High-voltage test techniques
                  Part 2 : Measuring systems ,1994
                  (DIN VDE 0432-2)
IEC Publ.270   Partial discharge measurements, 1981
                  (DIN VDE 0432-5 / 05.83)
IEC Publ.790   Oscilloscopes and peak voltmeters for impulse tests,
                  1984
                  (DIN VDE 0432-5 / 1985)
IEC Publ.1083-1   Digital recorders for measurements in high-voltage
                  impulse tests
                  Part 1: Requirements for digital recorders, 1991
                  (DIN VDE 0432-7 / 1992)
IEC Publ.1083-2   Digital recorders for measurements in high-voltage
                  impulse tests
                  Part 2: Digital signal processing, 1995
IEC Publ.833   Measurement of power frequency electric fields, 1987
IEC Publ.52   Recommendations for voltage measurement by means
                  of sphere-gaps (one sphere earthed), 1960.
                  (DIN VDE 0433-2)

References

AEG, Das Hochspannungsinstitut der AEG, Festschrift zur Eröffnung des Instituts in Kassel 1953, s. auch AEG Mitteilungen 43 (1953), S. 256-304.

Amsinck R., Schmidt W., Rozner W., Modern Gasinsulated AC-Testing Equipment for Factory and On-Site Testing up to 1 MV, 4. Int.Sym Hochspannungstechnik, Athen, 1983, 52.09.

Anderson J. C., Dielectrics, Chapmann and Hall, London, 1964.

Anis H., Trinh N. G., Train D., Generation of switching impulses using high-voltage testing transformers , IEEE Trans. Power Apparatus and Systems Vol. PAS-94 (1975), S. 187- 195.

Baatz H., Überspannungen in Energieversorgungsnetzen, Springer, Berlin, 1956.

Baldinger E., Kaskadengeneratoren, in *Flügge, S.* Handbuch der Physik, Bd. 44, Springer, Berlin 1959.

Bartnikas R., McMahon E. J., Engineering dielectric, Vol. I, Corona measurement and interpretation, ASTM, Philadelphia, USA, 1979.

Bertele H., Mitterauer J., Hochstromtechnik in der modernen Forschung und Entwicklung der Kernfusion, E. und M. 87 (1970), S. 139-152 und 305-353.

Bewley L. V., Travelling waves on Transmission Systems, 2. Aufl., Dover Publ., New York , 1951.

Beyer M., Boeck W., Möller K., Zaengl W., Hochspannungstechnik, Springer Verlag,1986.

Bishop M. J., Feinberg R., Grundsätzliche Verbesserung des Hochspannungs-Stoßgenerators , Anwendung des Polytrigatrons als Schaltgerät, E. und M., 88 (1971), Heft 2,S. 62-67.

Bishop M. J., Simon M. .F., The impulse generators at Les Renardieres, Trans IEEE PAS 91 (1972), 6, S. 2366-2376.

Boeck W., Eine Scheitelspannungs-Meßeinrichtung erhöhter Meßgenauigkeit mit digitaler Anzeige, ETZ-A, 84 (1963), H. 26, S. 883-886.

Boeck W., Developing a Standard for Testing GIS-Disconnector, IERE-Workshop on Gas- Insulated Substations, Toronto 1990, Paper 7-2.

Böning P., Das Messen hoher elektrischer Spannungen, Braun, Karlsruhe, 1953.

Böning P., Kleines Lehrbuch der elektrischen Festigkeit, Braun, Karlsruhe, 1955.

Carrara G., Zaffanella L., UHV Laboratory Clearance Tests, IEEE-Paper No. 68, P.692 PWR Chicago, 1968.

Craggs J. D., Meek J. M., High Voltage Laboratory Technique, Butterworth, London, 1954.

Deutsch F., Schalter für Hochstromimpulse bei hohen Spannungen, Bull. SEV, 55 (1964), Nr. 22,S. 1123-1129.

Dixon W. J., Massey F. J., Introduction to Statistical Analysis, McGraw Hill, New York, 1969.

Dommel H. W., Electromagnetic Transients Program Reference Manual (EMTP Theory Book), Bonneville Power Administration, 1986.

Duyan H., Hahnloser G., Traeger D., Pspice. Eine Einführung, Teubner Verlag, 1991.

Elsner R., Das neue Höchstspannungsprüffeld des Siemens-Schuckertwerke in Nürnberg, Siemens-Z., 26 (1952), H.6, S. 259-267.

Felici N. J., Elektrostatische Hochspannungsgeneratoren, Braun, Karlsruhe, 1957.

Feser K., Rodewald A., Eine triggerbare Mehrfachabschneidefunkenstrecke für hohe Blitz- und Schaltstoßspannungen, 1. Int. Symp. Hochspannungstechnik, München, 1972, 124-131.

Feser K., Probleme bei der Erzeugung hoher Schaltstoßspannungen im Prüffeld, Bulletin SEV, 65 (1974), Nr.7, S. 496-508.

Feser K., Transient Behaviour of Damped Capacitive Voltage Dividers of some Million Volts, IEEE Trans. on PAS, Vol. 93 ,1974, S. 116-121.

Feser K., Bemessung von Elektroden im UHV-Bereich, gezeigt am Beispiel von Toroid- elektroden für Spannungsteiler, etz A Band 96 (1975), Heft 4, S. 206-210.

Feser K., Auslegung von Stoßgeneratoren für die Blitzstoßpannungsprüfung von Transformatoren, Bulletin SEV, 69 (1978), Heft 18,S.973-979.

Feser K., Hochspannungsprüfungen von metallgekapselten, gasisolierten Anlagen Vor- Ort, Bulletin SEV, 72 (1981), 1, S. 19-26.

Feser K., Messung hoher Spannungen, etz, Bd. 104, 1983, H. 17, S. 881-887.

Feser K., Pfaff W., A potential free spherical sensor for the measurement of transient electric fields, IEEE Transactions on PAS-103, (1984), pp. 2904-2911.

Feser K., MIGUS-EMP-Simulator für die Überprüfung der EMV, etz Band 208, 1987, Heft 10,S. 420-423.

Feser K., Hughes R. C., Measurement of Direct Voltage by Rod-Rod Gaps, Electra, 117 (1988), . 23-24

Feser.K., Diagnostik für Isoliersysteme der elektrischen Energietechnik: Entwicklungstendenzen , ETG Fachbericht, 40 (1992), S. 143-154.

Fischer A., Hochspannungslaboratorien im In- und Ausland, ETZ-A, 90 (1969), H. 25, S. 656-662.

Fitch R. A., Howell V. T. S., Novel Principle of Transient High-Voltage Generation, Proc. IEE, 111(1964), Nr. 4, S. 849-855.

Flegler E., Einführung in der Hochspannungstechnik, Braun, Karlsruhe ,1964.

Frank H., Schrader W., Spiegelberg J., 3 MV AC Voltage Testing Equipment with Switching Voltage Extension - Technical Conception, First Operation and Results, 7. Internationales Symposium für Hochspannungstechnik, Dresden, 1991, Paper No. 52.04.

Früngel F., High Speed Pulse Technology, Vol. I u. II, Academic Press, New York,1965.

Gänger B., Der elektrische Durchschlag von Gasen, Springer, Berlin, 1953.

Gockenbach E., Measurement of Standard Switching Impulse Voltages by Means of Sphere-Gaps (One Sphere Earthed) , Electra, No. 136, 1991, S. 91-95.

Gontscharenko, G. M., Dmochwskaja, L. F., Shakov E. M., Ispitalelnie Ustanowki I ismeritelnie Ustroistwa W Laboratorijach wisokogo (Versuchsanlagen und Messeinrichtungen in Hochspannungslaboratorien), Naprijaschenija, Moskwa, 1966 .

Grabner K., 1400 kV Wechselspannungs-Prüfkaskade in Säulenbauweise, ELIN-Z., 19, (1967), S. 24-32.

Greenwood A., Electrical Transients in Power Systems, Wiley, New York, 1971.

Gsodam H., Stockreiter H., Das neu erbaute Hochspannungslaboratorium und Transformatorenprüffeld der ELIN-UNION im Werk Weiz, ELIN-Z., 17, (1965), 132-137.

Guenther A. H., Bettis J. R., The laser triggering of high voltage switches, Appl. Phys. Vol. 11, 1978, S. 1577-1613.

Hartig A., Unvollkommener und vollkommener Durchschlag in Schwefelhexafluorid, Diss. TH Braunschweig, 1966, (Beiheft Nr. 3 der ETZ).

Hayashi Ch., Non-linear Oscillations in Physical Systems, McGraw Hill, New York, 1964.

Hauschild W., Dietrich M., Schlufft W., Schwab H., Teller-Segmentelektroden für Hochspannungsprüfanlagen, Elektrie, 37, (1983), S. 189-192.

Hauschild W., Mosch W., Statistik für Elektrotechniker, VEB Verlag Technik, Berlin, 1984.

Hecht A., Elektrokeramik, Springer, Berlin, 1959.

Helmchen G., Die Entwicklung der Stoßspannungstechnik, ETZ-A, 84, (1963), H. 4, S. 107-113.

Heilbronner F., Das Durchzünden mehrstufiger Stoßgeneratoren, ETZ-A, 92 (1971), H. 6, S. 372-376.

Heise W., Tesla-Transformatoren, ETZ-A, 85 (1964), H.1, S. 1-8.

Heller B., Veverka A., Surge Phenomena in Electrical Machines, Akademia Prague, 1968.

Herb R. G., Van de Graaff Generators , in *Flügge,* S; Handbuch der Physik, Bd. 44, Springer, Berlin 1959.

Heyne V., Erweiterung des Trasformatorenprüffeldes und Neubau eines Hochspannungs- laboratoriums, BBC-Nachr., 51 (1969), H.2, S. 67-73.

v. Hippel A., Dielectric Materials and Applications, 2. Aufl. Wiley, New York, 1958.

Hövelmann F., Untersuchungen über das Stoßdurchschlagsverhalten von technischen Elektrodenanordnungen in Luft von Atmosphärendruck, Diss. T. H. Aachen, 1966.

Hylten-Cavallius N. R., Giao T. N., Floor Net Used as Ground Return in High-Voltage Test Areas, IEEE PAS 88 (1969), Nr. 7, S. 996-1005.

Hylten-Cavallius N. R., Calibration and Checking Methods of Rapid High-Voltage Impulse Measuring Circuits, IEEE PAS 89 (1970), Nr. 7, S. 1393-1403.

Hylten-Cavallius N. R., High Voltage Laboratory Planning, Emil Haefely & Cie. AG, Basel, 1986.

Imhof A., Hochspannungs-Isolierstoffe, Braun, Karlsruhe, 1957.

Jiggins A. H., Bevan J. S., Voltage Calibration of a 400 kV Van de Graaff Machine, J. Sc. Instruments, 43 (1966), S. 478-479.

Kachler A. J., Contribution to the problem of impulse voltage measurements by means of sphere gaps, 2. Int. Symp. Hochspannungstechnik, Zürich, 1975, S. 217-221.

Kaden H., Wirbelströme und Schirmung in der Nachrichtentechnik, 2. Aufl., Springer, Berlin, 1959.

Kärner H., Die Erzeugung steilster Stoßspannungen hoher Amplitude, Dissertation, TH München, 1967.

Kannan S. R., Narayana Rao Y., Prediction of the Parameters of an Impulse Generator for Transformer Testing, Proc. IEE, 120 (1973), 9, pp. 1001-1005.

Kappeler H., Hartpapierdurchführungen für Höchstspannung, Bull. SEV, 40 (1949), Nr. 21, S. 807-815.

Kieback D., Der Öldurchschlag bei Wechselspannung, Diss. TU Berlin, 1969.

Kind D., Meßgerät für hohe Spannungen mit umlaufenden Meßelektroden, ETZ-A,77 (1956), H. 1, S. 14-16.

Kind D., Die Aufbaufläche bei Stoßspannungsbeanspruchung technischer Elektrodenanordnungen in Luft, ETZ-A, 79 (1958),H. 3, S. 65-69.

Kind D., Salge J., Über die Erzeugung von Schaltspannungen mit Hochspannungs- Prüftransformatoren, ETZ-A, 86 (1965), H. 20, S. 648-651.

Kind D., König D., Untersuchungen an Epoxidharzprüflingen mit künstlichen Hohlräumen bei Wechselspannungsbeanspruchung, Elektrie, 21 (1967), S. 9-13.

Kind D., Empfehlungen für die Durchführung von Hochspannunsprüfungen von SF_6- Anlagen oder Rohrschienen Vor-Ort, etz-A, Heft 11, (1974), S. 588-589.

Kind D., Kärner H., Hochspannungs-Isoliertechnik, Vieweg, Braunschweig, 1982.

Kind D., Schon K., Schulte R., The calibration of standard impulse dividers, 6. Int. Symp. Hochspannungstechnik, New Orleans, 1989.

Knoepfel H., Pulsed High Magnetic Fields, North-Holland Publ. Co., Amsterdam, 1970.

Knudsen N. H., Abnormal Oscillations in Electric Circuits Containing Capacitance, Trans. of the Royal Institute of Technology, Nr. 69, Stockholm, 1953.

Kodoll W., Teilentladungsdurchschlag von polymeren Isolierstoffen bei Wechselspannung, Diss. TU Braunschweig, 1974.

Köhler W., Spannungsquellen für Fremdschichtprüfungen, Dissertation, Universität Stuttgart, 1988.

Knudsen, N.H., Abnormal Oscillations in Electric Circuits Containing Capacitance, Trans. of the Royal Institute of Technology, Nr.69, Stockholm 1953.

Kodoll, W., Teilentladungsdurchschlag von polymeren Isolierstoffen bei Wechselspannung, Diss. TU Braunschweig, 1974.

Kreuger E. H., Discharge Detection in High Voltage Equipment, Heywood, London, 1964.

Kreyszig E., Statistiche Methoden und ihre Anwendungen , Vandenhoeck & Ruprecht, Göttingen, 1967.

Küpfmüller K., Einführung in die theoretische Elektrotechnik, Springer, Berlin, 1965.

Kuffel E., Zaengl W. S., High-Voltage Engineering, Pergamon Press, Oxford, 1984.

Kuffel E., Zaengl W. S., High voltage engineering fundamentals, Pergamon Press, Oxford, 1984.

Läpple H., Das Versuchsfeld für Hochspannungstechnik des Schaltwerks der Siemens-Schuckert -Werke, Siemens Z. 40 (1966), H.5, S. 428-435.

Latzel H. G., Schon K., Precise Capacitance Measurements of High-Voltage Compressed Gas Capacitors, IEEE Transactions on Instruments and Measurements, Vol. IM-36, No. 2 (1987), S. 381-384.

Lautz G., Elektromagnetische Felder, Teubner, Stuttgart 1969.

Leroy G., Bouillard J. G., Galle G., Simon M., Essais di'electriques et tre's hautes tensions " Le L.T.H.T. des Renardie'res ", Socie'te' Francaise des Electriciens, 1e're Section, Avril 1971.

Leroy G., Gallet G., Ele'ments pour un projet de laboratoire a' haute tension, E.D.F. bulletin de la direction des e'tudes et recherches - serie B, re'seaux e'lectriques, mate'riels e'lectriques Nr. 3 / 4 (1975), S. 5-44.

Lesch G., Lehrbuch der Hochspannungstechnik, Springer, Berlin 1959.

Leschanz A., Oberdorfer G., Das neue Hochspannungsinstitut der Technischen Hochschule Graz, Elektrotechnik und Maschinenbau, 85 (1968), S. 527-532.

Leu J., Teilentladungen in Epoxydharz-Formstoff mit künstlichen Fehlstellen, ETZ-A, 87 (1966), S. 659-665.

Liebscher F., Held H., Kondensatoren, Springer, Berlin,1968.

Llewellyn-Jones F., Ionization and Breakdown in Gases, Science Paperbacks, London, 1957.

Lührmann H., Fremdfeldbeeinflussung kapazitiver Spannungsteiler, ETZ-A, 91 (1970), H.6, S. 332-335.

Lührmann H., Rasch veränderliche Vorgänge in räumlich ausgedhnten Hochspannungskreisen, Diss. TU Braunschweig, 1973.

Malewski R., Dechamplain A., Digital Impulse Recorder for High Voltage Laboratories, IEEE PAS, 99 (1980), S. 636-639.

Malewski R., Poulin B., Digital monitoring techniques for HV impulse tests, IEEE Trans. PAS 104, No. 11, 1985, S. 3108.

Marx E., Hochspannungspraktikum, 2. Aufl. Springer, Berlin, 1952.

Marx R., Zirpel R., Präzions-Meßeinrichtung zur Messung hoher Wechsel- und Gleichspannungen, PTB-Mitteilungen, 100-2 (1990), S. 119-124.

Matthes W., Zahorka R., Verzerrung der Spannungskurvenform von Prüftransformatoren infolge Oberschwingungen im Magnetisierungsstrom, ETZ-A, 80 (1959), S. 649-653.

Meek J. M., Craggs J. D., Electrical Breakdown of Gases, Clarendon Press, Oxford, 1953.

Micafil, Hochspannungs-Laboratorium Micafil, Firmenschrift zur Einweihung des Laboratoriums in Zürich, 1963.

Minkner R., Der Drahtwiderstand als Bauelement für die Hochspannungstechnik und Rechentechnik, Meßtechnik, 4 (1969), S. 101-106.

Modrusan M., Normierte Berechnung von Stoßstromkreisen für vogegebene Impulsströme, Bull. SEV, 67,1976, H. 22, S. 1232-1236.

Modrusan M., Langzeit-Stoßstromgenerator für die Ableiterprüfung gemäß CEI-Empfehlungen, Bull. SEV, 68, 1977, H. 24, S. 1304-1309.

Moeller J., Metal-Clad Test Transformer for SF_6-Insulated Switchgear, 2. Int. Symp. On High Voltage Engg. (1975), Paper No. 2.1-09, pp. 161-164.

Möller K., Spannungsabfälle in den Wänden metallisch abgeschirmter Hochspannungs-laboratorien, ETZ-A, 86 (1965), H. 13, S. 421-426.

Mole G., Basic Characteristics of Corona Detector Calibrators, IEEE PAS, 89 (1970), S. 198-204.

Mosch W., Die Nachbildung von Schaltüberspannungen in Höchstspannungsnetzen durch Prüfanlagen, Wiss. Z. TU Dresden, 18 (1969), H. 2, S. 513-517.

Müller, W., Untersuchungen der Spannungsform von Prüftransformatoren an einem Modell, Siemens-Zeitschrift, 35 (1961), S. 50-57.

Mürtz H., Hochspannungs-Explosionsverformung, ETZ-B, 16 (1964),H. 18, S. 529-535.

Nasser E., Heiszler M., Educational Laboratories in High-Voltage Engineering, IEEE- E 12, (1969), Nr. 1, S. 60-66.

Nasser E., Fundamentals of Gaseous Ionization and Plasma Electronics, Wiley, New York, 1971.

Oechsler F., Einsatzmöglichkeiten eines Teilchenbeschleunigers zur absoluten Messung hoher Gleichspannungen, Dissertation, Universität Karlsruhe, 1991.

Owen D. B., Handbook of Statistical Tables, Addison-Wesley, London, 1962.

Paasche P., Hochspannungs-Messungen, VEB Verlag Technik, Berlin, 1957.

Peier D., Stolle D., Ohmscher 1 MV-Teiler für Blitz- und Schaltstoßspannung, etz, Bd. 108, 1987, H. 6/7, S. 248-251.

Peier D., Elektromagnetische Verträglichkeit, Problemstellung und Lösungsansätze, Hüthig Verlag, Heidelberg, 1990.

Peschke E., Der Durch- und Überschlag bei hoher Gleichspannung in Luft, Dissertation, TH München, 1968.

Petersen C., Untersuchungen über die Zündverzugszeit von Dreielektroden-Funken-strecken, ETZ-A, 86 (1965), H. 17, S. 545-552.

Pfestorf G. K. M., Jayaram B. N., Über die theoretische Behandlung der Kaskaden- schaltung von Hochspannungstransformatoren, Jahrbuch der TH Hannover, 1958/1960.

Philippow, E., Nichtlineare Elektrotechnik, Akad. Verlagsanstalt, Leipzig, 1963.

Philippow E., Taschenbuch Elektrotechnik, Band 2, Starkstromtechnik, VEB Verlag Technik, Berlin, 1966.

Philippow E., Taschenbuch Elektrotechnik, Band 1, Grundlagen, VEB Verlag Technik, Berlin, 1968.

Potthof K., Widmann W., Meßtechnik der hohen Wechselspannungen, Vieweg, Braunschweig, 1965.

Prinz H., Hochspannungs-Messung mit dem rotierenden Voltmeter, ATM Blatt J 763-3,4,5 (1939).

Prinz H., Zaengl W., Ein 100 kV-Experimentierbaukasten, Elektizitätsw. 59 (1960), H. 20, S. 728-734.

Prinz H., Feuer, Blitz und Funke, Bruckmann, München, 1965.

Prinz H., Hochspannungsfelder, Oldenbourg, München, 1969.

Raether H., Electron Avalanches and Breakdown in Gases, Butterworth, London, 1964.

Rasquin W., Statistische Auswertung der Meßergebnisse von Durchschlag-Untersuchungen, Bull. SEV, 63 (1972), H. 5, S. 231-239.

Raupach F., MWB-Hochspannungslaboratorium, Firmenschrift zur Inbetriebnahme des Laboratoriums in Bamberg, 1969.

Reverey G., Verma M. P., Fremdschicht-Prüfverfahren und Untersuchung an verschmutzten Isolatoren im In- und Ausland, ETZ-A, 91 (1970), H.9, S. 481-488.

Rieder W., Plasma und Lichtbögen, Vieweg, Braunschweig, 1967.

Rizk F. A. M., Bourdage M., Influence of AC-Source Parameters on Flashover Characteristics of Polluted Insulators, IEEE Trans., Vol. PAS -104, 1985, pp. 948-958.

Rodewald A., Ausgleichsvorgänge in der Marxschen Verviel-fachungsschaltung nach der Zündung der ersten Schaltfunkenstrecke, Bull. SEV, 60 (1969), H. 2, S. 37-44.

Rodewald A., Feser K., The generation of lightning and switching impulse voltages in the UHV region with an improved Marx circuit, IEEE Transactions on PAS, Vol. 93 , 1974, pp. 414-417.

Roth A., Hochspannungstechnik, 4. Aufl., Springer, Wien, 1959.

Rüdenberg R., Elektrische Schaltvorgänge, 4. Aufl., Springer, Berlin, 1953.

Rüdenberg R., Elektrische Wanderwellen, 4. Aufl., Springer, Berlin, 1962.

Sachs L., Statistische Methoden - Ein Soforthelfer, Springer, Berlin, 1970.

Salge J., Peier D., Brilka R., Schneider D., Application of Inductive Energy Storage for the Production of Intense Magnetic Fields, Proc. 6th Symp. On Fusion Technology, Aachen, 1970.

Salge, J. Drahtexplosionen in induktiven Stromkreisen, Habilitationsschrift, TU Braunschweig, 1971.

Schiweck L., Untersuchungen über den Durchschlagsvorgang in Epoxydharz-Formstoff bei hohen Spannungen, ETZ-A, 90 (1969), H. 25, S. 675-678.

Schon K., Konzept der Impulsladungsmessung bei Teilentladungsprüfungen, etz Archiv, Band 8 (1986), Heft 9, S. 319-324.

Schuler R. H., Liptak G., A new method for the high voltage testing of field windings (interturn insulation) on large rotating electrical machines, CIGRE-Bericht No. 11-04, 1980, Paris.

Schwab A. J., Hochspannungsmeβtechnik, Springer, Berlin, 1969.

Siemens, Formel- und Tabellenbuch fuer Starkstrom-Ingenieure, 2. Aufl., Girardet, Essen, 1960.

Sirait T., Elektrische Ausgleichsvorgänge in den Erdflächenleitern von Hochspannungs-laboratorien, Diss. TH Braunschweig, 1967.

Sirotinski L.I., Hochspannungstechnik, Band I, Teil 1: Gasentladungen, VEB Verlag Technik, Berlin, 1955.

Sirotinski L.I., Hochspannungstechnik, Band I, Teil 2: Hochspannungsmessungen, Hochspannungslaboratorien, VEB Verlag Technik, Berlin, 1956.

Sirotinski L.I., Hochspannungstechnik: Äuβere Überspannungen, Wanderwellen, VEB Verlag Technik, Berlin, 1965.

Sirotinski L.I., Hochspannungstechnik: Innere Überspannungen, VEB Verlag Technik, Berlin, 1966.

Slamecka E., Prüfung von Hochspannungs-Leistungsschaltern, Springer, Berlin, 1966.

Stamm H., Porzel R., Elektronische Meβverfahren, VEB Verlag Technik, Berlin, 1969.

Stephanides H., Grundregeln fuer den Aufbau von Erdungssystemen in Hoch-spannungslaboratorien, E. und M., 76 (1959), S. 73-79.

Strigel R., Elektrische Stoβfestigkeit, Springer, Berlin, 1955.

Sun R., Feser K., Maier H. A., Balzer G., A simplified Method for the Determination of the Maximum Overvoltage Factor during Disconnector Operation in GIS, 7. ISH, Dresden, 1991, Paper No. 83.06.

Thione L., Kučera J., Weck K. H., Switching Impulse Generation Techniques Using High Voltage Testing Transformers, Electra, No. 43 (1975), S. 33-72.

Unger H. G., Theorie der Leitungen, Vieweg, Braunschweig, 1967.

Verma M. P., The Criterion for Pollution Flashover and its Application to Insulation Dimensioning and Control, CIGRE 1978, Paris, Paper No. 33-04.

Wehinger H., Ausgleichsvorgänge in Prüftransformatoren bei der Erzeugung von Schalt-stoßspannungen, Diss. TU Braunschweig, 1977.

Wellauer M., Einführung in die Hochspannungstechnik, Birkhäuser, Basel, 1954.

Widmann W., Stoßspannungs-Generatoren, ATM Blatt Z 44-6, 7, 8 (1962).

Wiesinger J., Einfluß der Frontdauer der Stoßspannung auf das Ansprechverhalten von Funkenstrecken, Bull. SEV, 57 (1966), Nr. 6, S. 243-246.

Wiesinger J., Funkenstrecken unter Blitz- und Schaltstoßspannungen, ETZ-A, 90 (1969), H. 17, S. 407-411.

Winkelnkemper H., Die Aufbauzeit der Vorentladungskanäle im homogenen Feld in Luft, ETZ-A, 86 (1965), H. 20, S. 657-663.

Wittler M., Peier D., Wirkung transienter elektromagnetischer Felder auf geschirmte Leiteranordnungen, EMV '90, Karlsruhe, VDE Verlag, 1990.

Whtehead, S., Dielectric Breakdown of Solids, Clarendon Press, Oxford, 1951.

Zaengl W., Völcker O., Messung des Scheitelwertes hoher Wechselspannungen, ATM Blatt V 3383-4, (1961).

Zaengl W., Das Messen hoher, rasch veränderlicher Stoßspannungen , Diss. TH München, 1964.

Zaengl W., Der Stoßspannungsteiler mit Zuleitung, Bull. SEV, 61 (1970), Nr. 21, S. 1003-1017.

Zaengl W., Bernasconi F., Bachmann B., Schmidt W., Spinnler K., Experience of a.c. voltage tests with variable frequency using a light weight on-site series resonant circuit, CIGRE-Session 1982, Paris, Paper No. 23-07.

Zischank W., Schutzfunkenstrecken zur Überspannungsbegrenzung bei direkten Blitzeinschlägen, Dissertation, Hochschule der Bundeswehr, München,1983.

Subject Index

Printed and bound by CPI Group (UK) Ltd, Croydon, CR0 4YY

03/10/2024

01040434-0020